# Introduction to Computational Physics for Undergraduates (Second Edition)

Online at: https://doi.org/10.1088/978-0-7503-6493-5

# Introduction to Computational Physics for Undergraduates (Second Edition)

**Omair Zubairi**
*Department of Physics, Florida Polytechnic University, Lakeland, FL, USA*

**Fridolin Weber**
*Department of Physics, San Diego State University, San Diego, CA, USA*

**IOP** Publishing, Bristol, UK

ISBN    978-0-7503-6493-5 (ebook)
ISBN    978-0-7503-6496-6 (print)
ISBN    978-0-7503-6495-9 (myPrint)
ISBN    978-0-7503-6494-2 (mobi)

DOI    10.1088/978-0-7503-6493-5

Version: 20241201

IOP ebooks

British Library Cataloguing-in-Publication Data: A catalogue record for this book is available from the British Library.

Published by IOP Publishing, wholly owned by The Institute of Physics, London

IOP Publishing, No.2 The Distillery, Glassfields, Avon Street, Bristol, BS2 0GR, UK

US Office: IOP Publishing, Inc., 190 North Independence Mall West, Suite 601, Philadelphia, PA 19106, USA

# Contents

# Preface to the second edition

This second edition of Introduction to Computational Physics for Undergraduates represents a substantial update, designed to meet the evolving needs of modern computational physics education. Based on modern Fortran, this book provides a rigorous foundation in coding and algorithm development, with explicit Python examples included to visually illustrate numerical data. The first edition was well-received, and in response to feedback from students and instructors, this new edition has been thoroughly revised and expanded.

These updates enhance the book's role as a practical resource for students aiming to build advanced skills in coding, computational techniques, and scientific data analysis.

The second edition introduces new coding exercises that emphasize real-world applications, giving students hands-on experience as they engage with complex physics problems. The exercises are designed to encourage active learning, guiding students progressively from fundamental concepts to more advanced computational methods. To better support students preparing for challenges in research and industry, many topics now include Python code examples that visualize computational results graphically.

All code examples and exercises are now provided in Jupyter Notebooks, allowing students to interact directly with the code, test various configurations, and deepen their understanding through experimentation.

These notebooks offer a flexible and accessible platform for beginners and advanced learners alike, enhancing the practical, hands-on experience for all students. In addition, the problem sets have been refined, featuring a broader selection of exercises closely aligned with course learning objectives from introductory tasks to more challenging projects.

Though designed for undergraduate physics students, this text is equally valuable to anyone interested in learning advanced coding techniques within a physics framework. By integrating computational tools with practical examples, this edition is well-suited for classroom use, self-study, or professionals in physics, engineering, or data science looking to expand their skill sets.

Our goal with this second edition is to bridge theoretical knowledge and practical application, equipping learners with essential computational skills for advanced study and professional work in scientific fields. We hope this updated edition inspires and empowers students to enhance their coding and analytical abilities, fostering a spirit of curiosity and innovation that will support their future in scientific research and beyond.

# Acknowledgements

F. Weber thanks the National Science Foundation (USA) for its support in creating valuable research and learning opportunities for undergraduate students. This funding has facilitated a variety of hands-on activities, allowing students to engage deeply with research topics in physics. The NSF's support was instrumental in helping students gain essential skills, build research experience, and foster a strong foundation for future scientific careers.

# Author biographies

## Omair Zubairi

**Omair Zubairi** received his BSc and MSc in Physics from San Diego State University. He obtained his PhD in Computational Science from Claremont Graduate University and San Diego State University where he primarily worked on compact star physics. His other research interests include general relativity, cosmology, numerical astrophysics and computational methods and techniques. Omair is a dedicated educator in physics, computational and data science. He has taught students, from all backgrounds in many areas of physics and computation from the introductory sequence to upper division courses where he incorporates numerical methods and computational techniques into each course. "By allowing students to see and apply numerical simulations to various topics covered in lectures, they are able to gain invaluable insight into the problem at hand".

## Fridolin Weber

**Fridolin Weber** is a Distinguished Professor of Physics at San Diego State University and a Research Scientist at the University of California, San Diego. His research focuses on nuclear and particle processes in extreme astrophysical systems, such as neutron stars and supernovae. His interests also include quantum many-body theory applied to nuclear and dense quark matter, relativistic astrophysics, quantum gravity, and Einstein's theory of general relativity. He has published five books, co-authored over 250 papers, and given over 300 talks at national and international conferences and physics schools.

**IOP** Publishing

# Introduction to Computational Physics for Undergraduates
## (Second Edition)

**Omair Zubairi and Fridolin Weber**

# Chapter 1

## The Linux/Unix operating system

## 1.1 Introduction

The main purpose of this introduction is to get you familiar with the interactive use of Unix/Linux for day-to-day organizational and programming tasks. Unix/Linux is an operating system (OS), which we can loosely define as a collection of programs (often called processes) that manage the resources of a computer for one or more users. These resources include the CPU, network facilities, terminal windows, file systems, disk drives and other mass-storage devices, printers, and many more. During the course, the most common way you will use Unix/Linux is through a command-line interface; you will type commands to create and manipulate files and directories, start up applications such as text editors or plotting packages, and compile and run Fortran programs.

When you type commands in Unix/Linux, you are actually interacting with the OS through a special program called a shell which provides a user-friendly command-line interface. These command-line interfaces provide powerful environments for software development and system maintenance. Though shells have many commands in common, each type has unique features. Over time, individual programmers come to prefer one type of shell over another. We recommend that you use the 'C shell' (csh), the 'Z shell' (zsh), or the 'Bash shell' (bash) for interactive use.

In choosing a command shell in a Unix-like environment, it is helpful to compare the key features and capabilities of Bash, Csh, and Zsh. As shown in table 1.1, Bash offers extensive scripting capabilities and compatibility with Bourne shell syntax, making it widely used in modern environments. Csh is more common in older Unix systems but offers fewer features for scripting. Zsh, on the other hand, combines the best of Bash and the KornShell (ksh) with additional enhancements. It provides

doi:10.1088/978-0-7503-6493-5ch1

advanced features and extensive customization options, making it increasingly popular among Unix/Linux users.

Glob patterns, or globbing, are used in Unix/Linux shells to specify sets of filenames with wildcard characters. They are commonly employed in Bash, Csh, and Zsh for pattern matching on file and directory names, as shown in tables 1.2 and 1.3. The asterisk * matches any number of characters, including none, while the question mark ? matches exactly one character. Square brackets [...] are used to match any one of the enclosed characters, and the patterns [!...] or [^...] match any one character not enclosed. Curly braces {...} can be used in Bash and Zsh to match any of the comma-separated patterns within, but this is not supported in Csh. The double asterisk ** is a powerful feature in Bash and Zsh that matches directories recursively, enabling pattern matching across multiple directory levels. While many glob patterns are supported similarly across Bash, Csh, and Zsh, some advanced features such as {...} and ** are not available in Csh. This provides users with flexible and efficient ways to handle file and directory operations through pattern matching in their scripts.

**Table 1.1.** Comparison of key features and capabilities of Bash, Csh, and Zsh shells.

| Feature | Bash | Csh | Zsh |
| --- | --- | --- | --- |
| Command-line Editing | Yes | Limited | Advanced |
| History Mechanism | Yes | Yes | Yes |
| Job Control | Yes | Basic | Advanced |
| Scripting Capabilities | Extensive | Basic | Advanced |
| Customization | Limited | Rigid syntax | Extensive |
| Tab Completion | Basic | Basic | Advanced |
| Prompt Customization | Yes | Basic | High |
| Globbing | Standardxs | Basic | Enhanced |
| Array Handling | Yes | No | Advanced |

**Table 1.2.** Summary of glob patterns in Bash, Csh, and Zsh.

| Pattern | Bash Usage |
| --- | --- |
| * | *.txt matches file.txt, document.txt |
| ? | file?.txt matches file1.txt, fileA.txt |
| [...] | file[123].txt matches file1.txt, file2.txt |
| [!...], [^...] | file[!123].txt matches file4.txt, fileA.txt |
| {...} | file{1,2,3}.txt matches file1.txt, file2.txt |
| ** | **/*.txt matches all .txt files in the directory and and subdirectories |

**Table 1.3.** This table details common glob patterns and their usage across three Unix/Linux shells: Bash, Csh, and Zsh. Note that while many patterns are similarly supported across these shells, some patterns such as {...} and ** are not supported in Csh.

| Pattern | Csh Usage | Zsh Usage |
|---|---|---|
| * | Same as Bash | Same as Bash |
| ? | Same as Bash | Same as Bash |
| [...] | Same as Bash | Same as Bash |
| [!...], [^...] | Same as Bash | Same as Bash |
| {...} | Not supported | Same as Bash |
| ** | Not supported | Same as Bash |

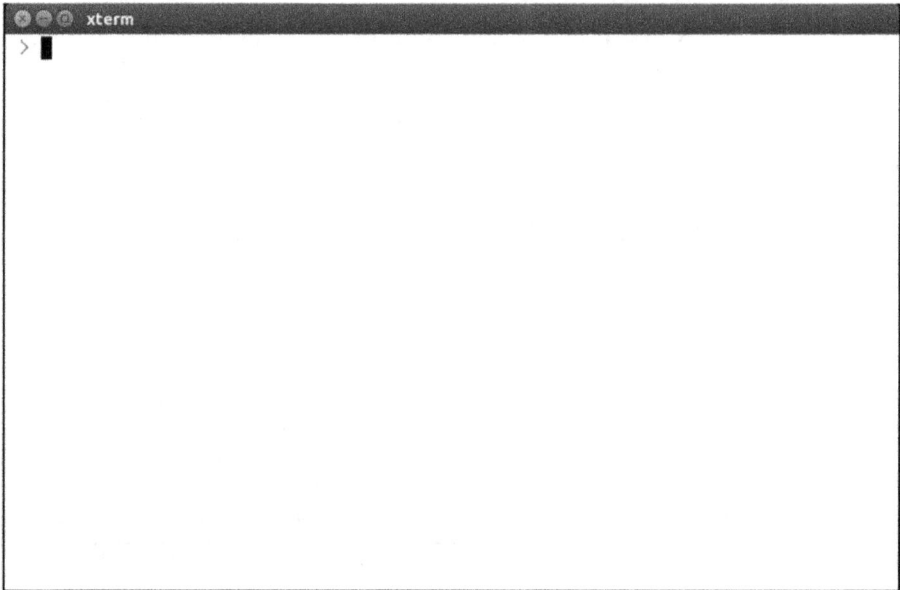

**Figure 1.1.** The X-terminal window on a machine running Ubuntu Linux. The shell prompt, shown as '>', indicates the shell is ready to accept commands; while a cursor, represented as a black rectangle, shows the current position for typing input.

All the Unix/Linux commands described below are Bash shell features. The Bash shell offers command-history recall and editing via the 'arrow' keys (as well as 'delete' and 'backspace'). After you have typed a few commands, hit the 'up arrow' key a few times and note how you scroll back through the commands you have previously issued. In the following, we shall assume that you have at least one active shell on each system in which to type Unix/Linux commands, and I will often refer to a window in which a shell is executing commands as the terminal. Popular terminal windows on Unix/Linux machines are iTerm, aterm, and xterm. An example of the latter is shown in figure 1.1. Henceforth, commands typed to the shell at the shell prompt (denoted ' ') are shown in red typewriter fonts, while the shell response is shown in blue typewrite fonts. Here is an example:

```
> pwd                                    ↵
/home/student
> whoami                                 ↵
student
> ps -p $$ -o comm=""                    ↵
-bash
```

## 1.2 Files and directories

There are essentially there types of files in Unix/Linux. These are:
- Regular files, such as plain text files, source code files, executables, or postscript files,
- Directory files, which contain other files and/or directories, and
- Special files, such as block files, character device files, named pine files, symbolic link files, and socket files.

### 1.2.1 Pathnames and working directories

All Unix/Linux file systems are rooted in a special directory called '/'. All files within the file system have absolute pathnames which begin with '/' and which describe the path down the file tree to the file in question. Thus,

/home/student/sample.txt

refers to a file named sample.txt which resides in a directory with absolute pathname

/home/student/

which itself lives in directory

/home

which is contained in the root directory, /. In addition to specifying the absolute pathname, files may be uniquely specified using relative pathnames. The shell maintains a notion of your current location in the directory hierarchy, known (appropriately enough) as the working directory. The name of the working directory may be printed using the pwd command:

```
> pwd                                                        ↵
/home/student/
```

If you refer to a filename such as

`file.txt`

or a pathname such as

`dir1/dir2/file.txt`

so that the reference does not begin with a '/', the reference is identical to an absolute pathname constructed by prepending the working directory followed by a '/' to the relative reference. Thus, assuming that you working directory is

`/home/student/txt`

the two previous relative pathnames are identical to the absolute pathnames

`/home/student/txt/file.txt`

`/home/student/txt/dir1/dir2/file.txt`

Note that although these files have the same filename, i.e., `file.txt`, they have different absolute pathnames and hence are different from each other.

Each user of a Unix/Linux system typically has a single directory, called the home directory, which serves as the base of their personal files. The command `cd` (change directory) with no arguments will always take you to your home directory. On your Linux machine you may see something like this:

```
> cd                                                         ↵
> pwd                                                        ↵
/home/student
```

When using the C-shell, you may refer to your home directory using a tilde ('~'). Thus, assuming the home directory is `/home/student`, then

```
> cd ~
```

followed by

```
> cd dir1/dir2
```

is identical to

```
> cd /home/student/dir1/dir2
```

Unix/Linux uses a single period ('.') and two periods ('..') to refer to the working directory and the parent of the working directory, respectively:

```
> cd ~/student/homework1                          ↩
> pwd                                             ↩
/home/student/homework1
> cd ..                                            ↩
> pwd                                             ↩
/home/student
> cd .                                             ↩
> pwd                                             ↩
/home/student
```

Note that

```
> cd .
```

does nothing—the working directory remains the same. However, the '.' notation is often used when copying or moving files into the working directory. See below for more details.

### 1.2.2 Filenames

There are relatively few restrictions on filenames in Unix/Linux. On most systems (including Linux machines), the length of a filename cannot exceed 255 characters. Any character except the forward slash ('/') and 'null' may be used. However, you should avoid using characters which are special to the shell, such as '(', ')', '*', '?', '$', '!', as well as blanks (spaces). In other words, using upper and lowercase letters, numbers, and a set of symbols, as shown below, is highly recommended,

a − z, A − Z, 0 − 9, _, ., −

which includes underscores, periods, and dashes. As is the case for other OSs, the period is often used to separate the 'body' of a filename from an 'extension'.

**Table 1.4.** Examples of file extensions.

| Full Filenames | Extensions |
| --- | --- |
| `program.f` | `.f` |
| `program.f90` | `.f90` |
| `paper.tex` | `.tex` |
| `document.txt` | `.txt` |

**Table 1.5.** Overview of standard Unix/Linux filename extensions.

| File Extensions | Usage |
| --- | --- |
| `.c` | C language source code |
| `.cpp` | C++ language source code |
| `.f` | Fortran 77 language source code |
| `.f90` | Fortran 90 language source code |
| `.o` | Object code generated by a compiler |
| `.pl` | Perl language source code |
| `.ps` | PostScript language source |
| `.tex` | TeX or LaTeX document |
| `.dvi` | Device independent output file |
| `.gif` | Graphic Interchange Format (GIF) graphics file |
| `.jpg` | Joint Photographic Experts Group (JPE) graphics file |
| `.tar` | Archive file created with `tar` |
| `.z` | Compressed file created with `compress` |
| `.tgz` | Compressed (gzipped) archive file created with `tar` |
| `.a` | Library archive file created with `ar` |

Examples of which are shown in table 1.4, where the full filenames are listed in the left-hand column and the extensions in the right-hand column. Note that unlike some other OSs, extensions are not required and are not restricted to some fixed length. Several standard Unix/Linux filename extensions are shown in table 1.5. The underscore and dash sign are often used to create more human readable filenames such as 'This_is_better', which easier to read than a file named 'Thisisnotsogood'.

If one accidentally creates a filename containing characters which are special to the shell, such as '*' or '?', it is best to rename or move (mv) this file. This is done by enclosing the file's name in single forward quotes to prevent shell evaluation. Below, we show an example for a text file which contains an asterisk:

```
> mv 'bad_file*_name.txt'    good_file_name.txt          ↵
```

The `mv` command renames the file specified on the command line. The single quotes must be forward-quotes because backward quotes have a completely different meaning to the shell.

## 1.3 Overview of Unix/Linux commands

Mastering Unix/Linux commands is crucial for efficiently navigating and manipulating Unix-like OSs. These commands form the backbone of system administration, software development, and everyday computing tasks in Unix/Linux environments. Understanding their functionality and how to effectively combine them empowers users to streamline workflows, automate processes, and troubleshoot issues effectively. Beginning Unix/Linux users are often overwhelmed by the number of commands they must learn in order to perform tasks. To assist such users, we discuss in this chapter the most commonly used Unix/Linux commands, which will allow users to perform many essential operations on Unix/Linux machines. An overview of the most important commands is provided in table 1.8. Commands such as `ls`, `cp`, `mv`, `rm`, and `mkdir` enable users to list directory contents, copy files, move files between directories, delete files and directories, and create new directories. Commands such as `cd` and `pwd` facilitate navigation through the directory structure, displaying the current working directory, and presenting a hierarchical view of directory contents. Commands such as `ps`, `df`, and `du` provide important insights into system processes, resource usage, disk space allocation, and file system usage. Finally, `tar` is a Unix command that creates and extracts collections of files called tarballs, which is useful for archiving and distributing multiple files while preserving file system attributes. The `zip` command compresses files into a single archive, reducing their size, and `unzip` extracts these files. The Secure Shell (SSH) protocol `ssh` enables secure connections to remote hosts, allowing command execution over encrypted networks. Finally, the Secure File Transfer Protocol (SFTP) uses SSH to securely transfer files between local and remote systems.

The general structure of Unix/Linux commands is schematically given by

```
command_name [options] [arguments]
```

where the square brackets may contain optional parameters. Options to Unix/Linux commands are frequently single alphanumeric characters preceded by a minus sign, as in this example:

```
> ls -l                                              ↩
> cp -R ...                                          ↩
> man -k ...                                         ↩
```

where the ellipses stand for directory names, respectively, commands which have been omitted. They are typically provided as arguments to shell commands, which do not start with a '-' symbol in front. Individual arguments are separated by whitespace, i.e., one or more spaces or tabs:

```
> cp file1 file2                                    ↵
> grep 'a string' file                              ↵
```

There are two arguments in both of the above examples. Note the use of single forward quotes needed when supplying the grep command with an argument (i.e., 'a string') which contains spaces. The command

```
> grep a string file
```

without quotes has three different arguments rather than just two, and thus has a completely different meaning.

### 1.3.1 Executables and paths

In Unix/Linux, a command such as ls or cp is usually a file, which is known to the system to be executable. To invoke the command, you must either type the absolute pathname of the executable file or ensure that the file can be found in one of the directories specified by your path. For the C-shell and Bash shell, the current list of directories which constitute your path is maintained in the shell variable, PATH. To display the contents of this variable, type

```
> echo $PATH                                        ↵
```

The '$' mechanism is the standard way of evaluating shell variables and environment variables alike. The resulting output generated by the C-shell may look something like this,

```
/home/student/bin:/home/student/local/bin:/usr/local/
sbin
```

The order in which path components (i.e., first /home/student/bin, then /home/student/local/bin, and then /usr/local/sbin) appear in the path is important. When you invoke a command without using an absolute pathname, such as

```
> ls
```

the system looks in each directory in your path, in the specified order, until it finds a file with the appropriate name. If no such file is found, the shell returns an error message. As an example, say you want to list all files and directories in a given directory. This is accomplished by typing `ls` at the shell prompt and hitting the return button. Instead of `ls`, however, say you erroneously type `list`, which does not exist on your machine. The shell therefore will return an error message such as

```
-bash:  list:   command not found
```

The path variable is typically set in your ~/`.login` file and/or preferably your ~/`.cshrc` or ~/`.bashrc` files, which reside in your home directory. Examining ~/`.cshrc` and ~/`.bashrc` you should see lines like

```
export PATH=/usr/local/bin:/home/student/bin:$PATH
```

```
set path=($path /usr/local/bin $HOME/bin)
```

for the Bash shell and the C-shell, respectively. These lines add the directories `/usr/local/bin` and `$HOME/bin` to the previous (system default) value of `PATH`. Also note the use of parentheses to assign a value containing whitespace to the shell variable. `HOME` is an environment variable which stores the name of the home directory. Thus,

```
set path=($path /usr/local/bin ~/bin)
```

will have the same effect as

```
set path=($path /usr/local/bin $HOME/bin)
```

**Control characters:** The control characters CTRL-D, CTRL-C, and CTRL-Z have special meanings or uses within a shell. Below we shall familiarize ourselves with the actions and typical usages of these control characters. We shall use a caret ('^') to denote the CTRL key. Then, for instance,

```
>  ^D
```

means pressing the (upper or lowercase) D-key while holding down the CTRL (control) key. If you try the above example, you will notice that the shell does not 'echo' the ^D. This is typical of control characters. When you type ^D, the OS sends all of the current lines that you have typed (but not the ^D itself) to the program (e.g., mail program, LATEX) doing the read, which may echo the characters EOT (end-of-transmission). Other commands such as `cat`, for instance, will not echo anything. In almost all cases, however, you should be presented with the shell prompt. By default, the C-shell and Bash shell exit when they encounter an EOF. So if you type ^D at a the shell prompt, the terminal will close automatically. This behavior can be changed by adding set ignoreeof to ~/.cshrc for the C-shell and export ignoreeof=1 to ~/.bashrc for the Bash shell.

The ^C interrupt kills (stop in a non-restartable fashion) commands (processes) which have been started from the command line of a terminal window. This is particularly useful for commands which are taking much longer to execute or producing much more output to the terminal than anticipated. Many commands catch interrupts and you may sometimes have to type more than one to stop the command.

The ^Z interrupt suspends, i.e., stops in a restartable fashion, commands which have been started from the shell. This is useful because it is often convenient to temporarily halt execution of a command.

### 1.3.2 Special files

The following files, all of which reside in your home directory, have special purposes and you should familiarize yourself with their content. The first one is .cshrc. Commands in this file are executed each time a new C-shell is started. The second file to note is .login. Commands in this file are executed after those in .cshrc and only for login shells. When interacting with Unix/Linux via a window system, it is easy to start an interactive shell which is not a login shell but for which you presumably want the same initialization procedures. Consequently, your .login should be kept as brief as possible and all your start-up commands should be put in .cshrc instead. Users of the Bash shell rather than the C-shell should put all their the start-up commands in .bashrc.

Note that files whose name begins with a period ('.') are called hidden files. They are not shown in a standard listing generated with ls, but can be printed by adding the -a operand to the listing command, as shown here:

```
> ls -a                                                    ↵
```

Listing the names of all files in your home directory is accomplished with

```
> cd ; ls -a                                               ↵
```

where we have introduced another piece of shell syntax, namely the ability to type multiple commands separated by semicolons (';') on a single line. If one wants to list only the hidden files and hidden directories in a given directory, the following command is to be executed:

```
> ls -d .*                                                    ↵
```

where the -d operand guarantees that directories are listed as plain files (not searched recursively) and the asterisk ('*') stands for any number of characters.

**Shell aliases:** The syntax of many Unix/Linux commands is quite complicated. Furthermore, the bare-bones version of some commands is less than ideal for interactive use, particularly by novices. The C-shell and Bash shell provide a mechanism called aliasing, which allows one to easily remedy these deficiencies in many cases. The basic syntax for aliasing is

```
alias name definition
```

where name is the name (use the same considerations for choosing alias names as for filenames, i.e., avoid using special characters) of the alias and definition tells the shell what to do when you type name at the shell prompt, as if it was a command. The following examples give a basic idea how this works. More details can be found in the system's man pages by typing man csh for the C-shell and man bash for the Bash shell. A convenient redefinition of the standard listing command, for instance, is

```
% alias ls 'ls -FC'          (for the C-shell)
> alias ls='ls -FC'          (for the Bash shell)
```

These aliases for the ls command uses the -F and -C options, which are described in the discussion of the ls command below. Note that single quotes in alias definitions are essential if the definitions contains white spaces. The commands

```
% alias rm 'rm -i'                                            ↵
% alias cp 'cp -i'                                            ↵
% alias mv 'mv -i'                                            ↵
```

define C-shell aliases for rm, cp, or mv, which will request confirmation before attempting to remove, copy, or move each file, respectively, regardless of the file's permissions. For the Bash shell, the above shell commands read

```
> alias rm='rm -i'                                            ↵
> alias cp='cp -i'                                            ↵
> alias mv='mv -i'                                            ↵
```

Making use of aliases is highly recommended for novices and experts alike. To see a list of all current aliases for a given shell, simply type

```
> alias                                                    ↵
```

Note that aliases defined interactively in a given shell exist only as long as the terminal session is open. To create aliases permanently, they need to be defined in ~/.aliases or ~/.bashrc, which are located in your home directory, or in profile.local, which resides in the /etc/ directory. The aliases are made available to shells with the source command, by typing

```
> source ~/.aliases                                        ↵
> source /etc/profile.local                                ↵
```

at the shell prompt. The source command tells the shell to execute the commands in the files supplied as arguments.

## 1.4 Basic commands

The following list is by no means exhaustive, but rather represents what I consider an essential base set of Unix/Linux commands with which you should familiarize yourself as soon as possible. Refer to the manual pages (see below) for additional information about these commands.

### 1.4.1 Getting help and information

Use man, which is short for manual, to display information about a specific Unix/Linux command. The -k option may be used in combination with man to display a list of commands which have something to do with a specific topic or keyword. For example, typing

```
> man -k xterm                                             ↵
```

returns all information found on the system about the X-terminal window. It cannot be overemphasized how important it is for users to become familiar with this command. Although the level of intelligibility for commands (especially for novices) varies widely, most basic commands are thoroughly described in the man pages, with usage examples in many cases. It helps to develop an ability to scan quickly through text looking for specific information you might feel to be of use. Typical usage examples include:

```
> man man
```

to get detailed information on the man command itself,

```
> man cp
```

for information on cp, and

```
> man -k 'working directory'
```

to get a list of commands having something to do with the topic working directory. The command apropos, found on most Unix/Linux systems, is essentially an alias for man -k.

### 1.4.2 Communicating with other computers

The OpenSSH secure shell client ssh, a remote login program, can be used to securely login to another computers on the Internet and perform command-line operations on them interactively. These computers could be physically located anywhere in the world. ssh is the most common way to access remote Linux and Unix-like machines. The typical usage of ssh is either

```
> ssh remote.host.name -l login_name                    ↵
```

or, alternatively,

```
> ssh login_name@remote.host.name                       ↵
```

which initiates the login of a user named login_name on the remote machine with the network ID remote.host.name. The -l option specifies the login name of the user on the remote machine. Let us look at an example. As login_name we pick student. The login session is then initiated by typing ssh student@remote.host.name at the shell prompt (>) of the local machine, as shown below:

```
> ssh student@remote.host.name          (on the local machine)

student@remote.host.name's password:  xxxxxx         ↵
Login successfull from remote.host.name
student@remote >
```

If user student is known on the remote machine, they will be asked for the password. Hereupon, the remote machine returns the shell prompt, which allows student to run programs on the remote machine. If the login attempt fails because of a wrong password or an incorrect user name, a permission denied message will be printed and the failed login attempt will most likely be recorded on the remote machine. In the above example, commands processed at the local machine are shown in red typewriter font, while those processed at the remote machine are shown in black typewriter font. To leave the remote terminal window session, type exit at the shell prompt.

SFTP enables secure file transfer capabilities between networked machines. It also provides remote file system management functionality, allowing users to list the contents of remote directories and to delete remote files. Below is an example which illustrates how SFTP is used to copy a file named thesis.pdf from the remote host remote.host.name to the local host local.host.name. The user name is again a student who has an account on the remote host:

```
> sftp student@remote.host.name              (on the local machine)

student@remote.host.name's password:  xxxxxx        ↩
sftp> pwd                                            ↩
Remote working directory:  /home/student
sftp> ls                                             ↩
public     temporary     numerical_codes     Thesis
sftp> cd Thesis                                      ↩
sftp> pwd                                            ↩
Remote working directory:  /home/student/Thesis
sftp> ls                                             ↩
thesis.pdf
sftp> get thesis.pdf                                 ↩
Fetching /home/student/Thesis/thesis.pdf to thesis.pdf
/home/student/Thesis/thesis.pdf 100% 910KB 500.6KB/s
sftp> !ls                                            ↩
thesis.pdf
sftp> quit                                           ↩
```

As in the previous example, the command shown in red is typed on the local machine, and the commands and messages on the remote machine are in black. The commands ls, cd, and pwd are used to, respectively, list the files and directories, change directories, and print the name of the current directory on the remote machine. The transfer of a file, thesis.pdf in the current example, from the remote machine to the local machine is accomplished with the `get` command. Conversely, the command `put` should be used if a file is sent from the local machine to the remote machine. The command `!ls` is used to list the files and directories on the local machine, without leaving the SFTP session. Similarly `lcd` and `lpwd` can be used to change the working directory and to display the current working directory on the local server. Submitting the `quit` command terminates the SFTP session, as shown in the example above.

The `sftp` program offers a fairly extensive online help, which can be retrieved by typing

```
sftp> help                                                ↩
```

or by submitting one of the following commands:

```
sftp> help bin                                            ↩
sftp> help cd                                             ↩
sftp> help lcd                                            ↩
sftp> help put                                            ↩
sftp> help get                                            ↩
sftp> help prompt                                         ↩
sftp> help mget                                           ↩
```

### 1.4.3 Creating, manipulating, and viewing files and directories

The text editors which will be considered in this book are 'vi' and 'Emacs'. The vi editor (short for visual editor) is a simple screen editor which is available on almost all Unix systems. Emacs belongs to a family of text editors that are characterized by their enormous extensibility. Either of these two editors is perfectly suited to create, modify, and view text files at the level required for this course. Both editors are very popular among programmers, scientists, engineers, students, as well as system administrators. A brief introduction to vi and Emacs is provided in chapter 2. Most often `vi`, or its improved version named `vim`, is started to edit a single file with the commands

```
> vi filename                                             ↩
> vim filename                                            ↩
```

Similarly, the command to start an Emacs session at the shell prompt is given by

```
> emacs filename                                                    ↵
```

The command 'more' is used to view the contents of one or more files one page at a time. For example, executing the more command as

```
> more ~/.bashrc                                                   ↵
## Source global definitions
if [ -f /etc/bashrc ]; then
.   /etc/bashrc
fi

## Source local definitions
if [ -f /etc/profile.local ]; then
.   /etc/profile.local
fi

alias rm='rm -i'
alias mv='mv -i'
alias cp='cp -i'
alias dir='ls -aF'
```

displays the first page of lines of the ..bashrc configuration file, which is located in the home directory. The next page of lines (if any) is displayed by hitting the spacebar. Scrolling backward by one page is accomplished by typing b. Forward scrolling by one page is done by typing d | and the command q quits viewing a file. Consult the man pages (man  more) for the many other features of the more command.

The commands lp or lpr are used to print files. By default, files are sent to the system default printer, or to the printer specified in your **PRINTER** environment variable. The typical usage is

```
> lp -d laser print.ps                                             ↵
```

which prints postscript file print.ps at the printer named laser. If you want to print a regular text file or the source code of a numerical program such as Fortran or C++, it is highly recommended to convert these files first to postscript files using the enscript command. The typical usage is

```
enscript -o print.ps file.txt                          ←
enscript -o print.ps file.f90                          ←
enscript -o print.ps file.cpp                          ←
```

For detailed information about this command, type `man enscript`.

The commands `cd` and `pwd` are used to, respectively, change and display the current working directory. Below, we show a summary of these typically used commands. Note the usage of semicolons to separate distinct Unix/Linux commands issued on the same line:

```
> cd                                                   ←
> pwd                                                  ←
/home/student
> cd ~; pwd                                            ←
/home/fweber
> cd /tmp; pwd                                         ←
/tmp
> cd ..; pwd                                           ←
/
```

Recall that '..' refers to the parent directory of the working directory, so that

```
> cd ..                                                ←
```

takes you up one level in the file system hierarchy.

The listing command `ls` is used to list the contents of one or more directories, as shown for the home directory in this example:

```
> cd                                                   ←
> ls                                                   ←
Desktop Downloads thesis numerical homework paper.pdf
```

The listing can be made more explicitly by redefining the `ls` command as

```
> alias ls='ls -F'                                              ↩
```

which causes ls to append special characters, notably '*' for executables, '@' for links, and '/' for directories, to the names of certain files and directories. Then,

```
> ls                                                            ↩
Desktop/ Downloads/ thesis/ numerical/ homework/ paper.pdf
```

which immediately reveals that Desktop, Downloads, thesis, numerical, and homework are directories and paper.pdf is a regular file. To display hidden files in directories, the -a option is to be used:

```
> cd ~; ls -aF                                                  ↩
.bashrc .bash_profile .local/ .profile .vim .xemacs
Desktop/ Downloads/ thesis/ numerical/ homework/
paper.pdf
```

Finally, using ls in combination with the -l option allows one to display file and directory information in long format:

```
> cd  /numerical; ls -lF                                        ↩
-rwxr-x-- 15 student users 409 Aug 21 19:18 data
lrwxrwxrwx 11 student users 817 Mar 22 14:19 f77@ -> bu/
drwxr-xr-x 51 student users 170 Apr 20 12:34 f90/
drwxr-xr-x 17 student users 130 Aug 20 13:03 f2008/
drwxr-xr-x 70 student users 990 Feb 22 23:51 cpp/
```

The output in this case is worthy of a bit of explanation. First, observe that ls produces one line of output per file and directory listed. The first field in each listing consists of 10 characters (i.e., letters and dashes), which are further subdivided as follows:

- The first character is either 'a' if the listing refers to a regular file, a 'd' for a directory, or an 'l' for a link.
- The next nine characters refer to three groups (i.e., user, group, and other or world) of three characters each specifying read (r), write (w), and execute (x) permissions for the user (owner of the file), users in the owner's group, and all

other users. A '-' in the permission field indicates that the particular permission is denied.

Thus, in the above example, data is a regular file, with read, write, and execute permissions enabled for the owner (user student); read and execute permissions enabled for the members belonging to group users; and read, write, and execute denied for all other users. Note that you must have execute as well as read permissions for a directory in order to be able to change (cd) to this directory. See chmod below for more information on setting file permissions. Continuing to decipher the file listing, the next column in the above example lists the number of links to this file, then comes the name of the user who owns the file and the owner's group. This is followed by the size of the file in bytes, the date and time the file was last modified, and finally the name of the file. If any of the arguments to ls is a directory, then the contents of the directory is listed. Finally, we note that the -R option is used to recursively list subdirectories encountered in a given directory

```
> cd ~; pwd; ls                                          ↵
/home/student
Desktop/ Downloads/ thesis/ numerical/ homework/
paper.pdf
> ls -R /homework                                         ↵
instructions.txt
assignment.tex

homework//HW1:
code.f90
code.f90.ps
figures.ps
```

The command mkdir is used to make (create) directories. The following example illustrates how to create a directory named tempdir in a user's home directory:

```
> cd ~                                                   ↵
> mkdir tempdir                                          ↵
> cd tempdir; pwd                                        ↵
/home/student/tempdir
```

If one wants to create a deep directory, i.e., a directory for which one or more parent directories do not exist, the -p option is to be used in combination with mkdir, which automatically creates parent directories when needed:

```
> cd ~                                                              ↵
> mkdir -p dir1/dir2/dir3/dir4                                      ↵
> cd dir1/dir2/dir3/dir4; pwd                                       ↵
/home/student/dir1/dir2/dir3/dir4
```

In this case, the mkdir command creates the ~/dir1 directory first, followed successively by ~/dir1, ~/dir1/dir2, ~/dir1/dir2/dir3, and finally ~/dir1/dir2/dir3/dir4, all residing in the home directory, ~, of user student.

Copying files is accomplished with the cp command. This command can be used to create an identical copy of a file, copy one or more files to different directories, or duplicate an entire directory structure. The simplest usage is

```
> cp file1 file2                                                   ↵
```

which copies the contents of file1 to file2 in the current working directory. Assuming that cp is aliased to cp -i, which is highly recommended, the aliased command returns a prompt to the terminal window before a file will be copied that would overwrite an existing file. If the user's response from the terminal is 'y', then the file copy is carried out. Typing 'n' cancels the file copy. Below is an example of how this works. Assuming that file2 already exists in the current working directory, the shell dialog may be as follows:

```
> cp -i file1 file2                                                ↵
overwrite file2?  (y/n [n])                                        ↵
not overwritten
```

For many systems, [n] is the default option, as shown above, in which case only a simple shell return (↵) is required to not overwrite file2. To copy one or more files to a different directory, the typical command usage is

```
> cp -i file1 file2 temporary/.                    ↩
```

which attempts to copy file1 and file2 to sub-directory temporary. If files with identical names already exist in temporary, prompts at the terminal window will be returned which allow the user to either overwrite or cancel the file copy. An example which overwrites existing files is shown here:

```
> cp -i file1 file2 temporary/.                    ↩
overwrite temporary/./file1?  (y/n [n]) y          ↩
file1 → temporary/./file1
overwrite temporary/./file2?  (y/n [n]) y          ↩
file2 → temporary/./file2
```

Finally, duplicating an entire directory structure is done by adding the '-r' (recursive) option to cp. This copies the entire directory hierarchy to a new directory, such as

```
> cp -ir temporary/.  copy_temporary             ↩
```

The command mv is used to rename files or to move files from one directory to another. Again, let us assume that mv is aliased to mv -i so that the user will be prompted if an existing file would be clobbered by the move command. Here is an example illustrating the usage of the mv command:

```
> ls                                              ↩
fileA
> mv fileA fileB                                  ↩
> ls                                              ↩
fileB
```

The following sequence of commands illustrates how files file1, file2, and file2 located in /home/student/subdir1/subdir2/subdir3 can be moved up one level in the directory structure:

```
> pwd                                                      ↵
/home/student/subdir1/subdir2
> ls                                                       ↵
subdir3
> cd subdir3                                               ↵
> ls                                                       ↵
file1 file2 file3 file4
> mv file1 file2 file3 ../.                                ↵
> ls                                                       ↵
file4
> cd ..                                                    ↵
> pwd                                                      ↵
/home/student/subdir1/subdir2
> ls                                                       ↵
file1 file2 file3 subdir3
```

The rm command is used to remove (delete) files or directory hierarchies. The use of the alias rm -i, which requests confirmation (y for yes, n for no) before attempting to remove files or directories, is highly recommended. Note that once a file or directory has been removed in Unix/Linux there is essentially nothing you can do to restore them other than restoring a copy from a backup. In this example

```
> rm -i oldStuff.dat                                       ↵
remove oldStuff.dat?  y                                    ↵
oldStuff.dat
```

the remove command is used to delete a data file named oldStuff.dat once the action is confirmed with yes, the command

```
> rm file1 file2 file3                                     ↵
```

removes several files at once, and

```
> rm -r thisDir                                            ↵
```

removes the entire content of directory thisDir, including the directory itself. Be particularly careful when using the -r option because all files and directories will be irrevocably lost. Using the remove command without the -r option, i.e., > rm thisDir, cannot be used to remove thisDir. If submitted at the shell prompt, Unix/Linux will complain that thisDir is a directory and no action will be taken.

The command chmod is used to modify the permissions (file mode bits) of file. See the discussion of ls above for a brief introduction to file permissions and check the man pages for ls and chmod for additional information. Basically, file permissions control who can do what with the files. This includes yourself (the user, u), users in your group (g), and the rest of the world (the others, o). The file mode bits include the read bit (r), write bit (w), and the execute bit (x). When a user creates a new file, the system sets the permissions (mode bits) of a file to default values, which can be modified with the umask command (see man umask for more information). The default umask on many Unix/Linux systems is 022, which means that newly created files are readable by everyone (i.e., the world), but only writable by the owner, as shown below:

```
> touch newFile                                          ↵
> ls -dl newFile                                         ↵
> -rw-r-r- 1 student users 0 Aug 25 12:55 newFile
```

To change the umask setting of the current shell to something else, say 077, run

```
> umask 077                                              ↵
```

which changes the file mode bits for any newly created file in that shell to rw————. On Unix/Linux machines, the defaults should be such that you can do anything you want to a file you have created, while the rest of the world (including fellow group members) normally has only read and, where appropriate, execute permission. As the man page will tell you, you can either specify permissions in numeric (octal) form or symbolically. The latter are more intuitive and easier to remember. Several useful examples are shown below. Let us begin with

```
> chmod go-rwx file.f90                                  ↵
```

which removes all permissions from group and others. A file listing is therefore produced on the following terminal window output,

```
> ls -dl file.f90                                        ↵
> -rw---- 1 student users 33 Aug 13:09 file.f90
```

To make a file executable by everyone, the a option, which stands for all (i.e., user, group, and other), can be used,

```
> chmod a+x file.o                                    ↵
```

To remove this permission from everyone, the a-x option would be used. Finally, as a last example, the command

```
> chmod u-w thesis.tex                                ↵
```

removes the user's write permission to a file to prevent accidental modification of particularly valuable information, such as a thesis. As indicated above, file permissions are granted by putting a '+' sign after ugo or a and are removed by putting a '-' sign there.

## 1.5 More on the C-shell

### 1.5.1 Shell variables

The C-shell (csh) maintains a list of local variables, some of which, such as path, term, and shell, are always defined and serve specific purposes within the shell. Other variables, such as filec and ignoreeof, are optionally defined and frequently control details of shell operation. Finally, you are free to define you own shell variables as you see fit, but beware of redefining existing variables. By convention, shell variables have all-lowercase names. To see a list of all currently defined C-shell variables, simply type

```
% set                                                 ↵
```

or

```
% set | more                                          ↵
```

at the C-shell prompt (%). Using more will display as many lines as fit on the screen and prompts the shell to wait for user input (i.e., ↵) to advance. To print the value of a particular variable, use the Unix/Linux echo command plus the fact that a $ symbol in front of a variable name causes the evaluation of that variable,

```
% echo $PATH                                                    ←
```

To set the value of a shell variable use one of the following two ways,

```
% set thisvar=thisvalue                                         ←
% echo $thisvar                                                 ←
thisvalue
```

or

```
% set thisvarlist=(value1 value2 value3)                        ←
% echo $thisvarlist                                             ←
value1 value2 value3
```

Shell variables may be defined without being associated a specific value, as shown here:

```
% set somevar                                                   ←
% echo $somevar                                                 ←
```

The shell frequently uses this 'defined' mechanism to control enabling of certain features. To undefine a shell variable use unset, as in

```
% unset somevar                                                 ←
% echo $somevar                                                 ←
somevar:  Undefined variable.
```

A list of some of the main shell variables (predefined and optional) and their functions follows:
- path: Stores the current path for resolving commands.
- prompt: The current shell prompt—what the shell displays when it is expecting input.
- cwd: Contains the name of the (current) working directory.

- term: Defines the terminal window type. If your terminal is acting strangely, the command

```
% set term=vt100; resize          ↵
COLUMNS=87;
LINES=23;
export COLUMNS LINES;
```

often provides a quick fix.
- noclobber: When set, prevents existing files from being overwritten via output redirection (see below).
- filec: When set, this enables file auto completion. Partially typing a filename, using an initial sequence which is unique among files in the working directory, followed by hitting the TAB button will result in the system doing the rest of the typing of the filename for you.
- shell: Defines which particular shell you are using.
- ignoreeof: When set, this will disable shell-logout when ^D is typed.

### 1.5.2 Environment variables

Aside from the shell variables discussed in section 1.5.1, Unix/Linux systems utilize another type of variable known as environment variables, which are typically used for communication between the shell (not necessarily the C-shell) and other processes. By convention, environment variables are named using all uppercase letters. To effectively configure your shell environment, it is crucial to understand the differences in environmental settings among Bash, Csh, and Zsh. Table 1.6 provides a detailed overview of the environmental settings specific to Bash, including how to set and unset variables, default startup files, and other key features. For a comparison of similar settings in Csh and Zsh, refer to table 1.7.

**Table 1.6.** Environmental settings in Bash.

| Environmental Setting | Bash |
| --- | --- |
| Setting variables | export VAR=value |
| Unsetting variables | unset VAR |
| Default startup files | ~/.bashrc, ~/.bash_profile ~/.profile |
| Path variable | PATH |
| Alias creation | alias name='command' |
| Function definition | name() { commands; } |
| Command history file | ~/.bash_history |
| Command history variable | HISTFILE, HISTSIZE, HISTFILESIZE |
| Prompt customization | PS1 |
| Sourcing files | source file or . file |

**Table 1.7.** Environmental settings in Csh and Zsh.

| Environmental Setting | Csh | Zsh |
|---|---|---|
| Setting variables | `setenv VAR value` | `export VAR=value` |
| Unsetting variables | `unsetenv VAR` | `unset VAR` |
| Default startup files | `~/.cshrc`, `~/.login` | `~/.zshrc`, `~/.zprofile`, `~/.zshenv`<br>`~/.zlogin`, `~/.zlogout` |
| Path variable | `path` (array) | `PATH` |
| Alias creation | `alias name 'command'` | `alias name='command'` |
| Function definition | `alias name 'commands'` | `name() { commands; }` |
| Command history file | `~/.history` | `~/.zsh_history` |
| Command history variable | `history`, `savehist` | `HISTFILE`, `HISTSIZE`, `SAVEHIST` |
| Prompt customization | `prompt` or `set prompt` | `PROMPT`, `RPROMPT` |
| Sourcing files | `source file` | `source file` or `. file` |

**Table 1.8.** Summary of essential Unix/Linux commands.

| Topic | Command | Examples |
|---|---|---|
| List filenames | `ls` | `ls -l *.c`, `ls -a`, `ls -F`, `ls -alF` |
| Move files or directories | `mv` | `mv temp.txt newfile.txt` |
| Copy files or directories | `cp` | `cp temp.txt newfile.txt` |
| Remove files and directories: | `rm` | `rm temp.txt` |
|   delete files 'interactively' | | `rm -i temp.txt` |
|   delete files and directories 'forcefully' | | `rm -rf directory` |
| Display the user manual | `man` | `man rm` |
| Make a directory | `mkdir` | `mkdir newdir` |
| Remove a directory | `rmdir` | `rmdir newdir` |
| Change directory | `cd` | `cd texdir` |
| Print working directory | `pwd` | `pwd` |
| Send file to a printer | `lpr`, `lp` | `lp -d printer filename` |
| View contents of a file: | | |
|   display entire content | `cat` | `cat file` |
|   one screenfull at a time | `more` | `more file` |
|   with advanced features | `less` | `less file` |
| Print string or variable | `echo` | `echo "hello, world"` |

| | | |
|---|---|---|
| To see list of recent commands | `history` | `history` |
| Set protection of a file | `chmod` | `chmod 755 file` |
| Set owner of a file | `chown` | `chown smith file` |
| Make a link (alias) to a path | `ln` | `ln -s ~/classes/comp comp` |
| Find out disk quota | `quota` | `quota -v` |
| Find out disk usage | `du` | `du` |
| Display free disk space | `df` | `df -h` |
| Display process status | `ps` | `ps -j` |
| Create archive file 'bup.tar' containing all files and directories in 'project' | `tar -cvf` | `tar -cvf bup.tar project` |
| Create gzipped archive file of 'project' | `tar -czvf` | `tar -czvf bup.tar.tgz project` |
| Extract contents from an archive file | `tar -xvf` | `tar -xvf bup.tar` |
| Extract contents from gzipped archive file | `tar -xzvf` | `tar -xzvf bup.tar.gz` |
| List the contents of an archive file | `tar -tvf` | `tar -tvf bup.tar` |
| List the contents of a gzipped archive file | `tar -tzvf` | `tar -tzvf bup.tar` |
| Create a ZIP archive | `gzip` | `gzip -r bup.zip project` |
| Extract contents of a ZIP file | `gunzip` | `gunzip bup.zip` |
| Using 'bzip2' single-file compressor | `bzip2` | `bzip2 file.dat` |
| Decompressing 'bz2' files | `bunzip2` | `bunzip2 file.dat.bz2` |
| Convert text files to PostScript | `enscript` | `enscript -o code.ps code.f90` |
| Convert standard input to PostScript | `a2ps` | `a2ps -o code.ps code.f90` |
| Convert PostScript file to PDF file | `ps2pdf` | `ps2pdf code.ps code.pdf` |
| Printing and pagination filter for text files | `pr` | `pr code.f90 > code.f90.pr` |
| Secure communication with other hosts: 'ssh' (Secure Shell) | `ssh` | `ssh user@host` |
| Transferring files between hosts: 'scp' (secure file copy) | `scp` | `scp user@host:file file` |
| 'sftp' (secure file transfer protocol) | `sftp` | `sftp user@host` |

In the C-shell, you can display the value of all currently defined environment variables by typing

```
% env | more                                          ↵
```

Some environment variables, such as PATH, are automatically derived from shell variables. Others have their values set, typically in ~/.cshrc or ~/.login, using the syntax

```
% setenv VARNAME value                                    ↵
```

Note that, unlike the case of shell variables and set, there is no '=' sign in the assignment. The values of individual environment variables may be displayed using the commands printenv or echo:

```
% printenv HOME                                           ↵
/home/student
% echo $HOME                                              ↵
/home/student
```

It should be noted that, as with shell variables, the '$' sign causes the evaluation of an environment variable. It is particularly notable that the values of environment variables defined in one shell are inherited by commands (including C and Fortran programs, and other shells) which are initiated from that shell. For this reason, environment variables are widely used to communicate information to Unix/Linux commands (applications). The DISPLAY environment variable is a canonical example of an environment variable. It tells X-applications which display (screen) to use for output. It is typically set on remote machines so that output appears on the local screen. For example, assuming you are remotely logged into host darwin from the console of your local machine magic, then at the darwin prompt you may want to type

```
% setenv DISPLAY darwin:0.0                               ↵
```

after which all X-applications started on darwin will be displayed graphically on the local magic machine. If you encounter problems transporting windows from a remote machine to your local console, try typing xhost + at a shell prompt on the local machine. See man xhost for more information.

The HOME variable asks the shell to substitute the environment variable HOME. For example,

```
% cd $HOME/homework                                    ↵
```

allows you to change from any (sub) directory directly to homework, provided it exists in your home directory. Since HOME stand for your home directory ( / ), this command is equivalent to

```
% cd ~/homework                                        ↵
```

The PRINTER variable defines the default printer for use with lpr, lp, or programs such as enscript, which feed postscript files to a printer via lpr or lp. The default printers may be designated in ~/.cshrc by adding setenv PRINTER printer-name, or in ~/.bashrc by adding PRINTER=printername; export PRINTER, which sets the PRINTER variable for the C-shell and Bash shell, respectively.

### 1.5.3 C-shell pattern matching

The C-shell provides facilities which allow you to concisely refer to one or more files whose names match a given pattern. The process of translating patterns to actual filenames or pathnames is known as filename, respectively, pathname expansion, or globbing. The name expansions expands the '*', '?', and a pattern list ' [...] ' when you type them as part of a command. For example,

```
% *.ps                                                 ↵
```

lists all postscript files in a given directory, where the '*' acts as a placeholder for any string of characters. Replacing the asterisk with a question mark in the above command, i.e.,

```
% ?.ps                                                 ↵
```

causes the shell to list all postscript files with only one-character filenames, such as a.ps or 2.ps. Pattern lists [...] are constructed using plain text strings sandwiched between square brackets, such as

```
% ls [A-Z]*.ps                                         ↵
```

which lists all postscript files that start with any capital letter. If desired, these files could then be moved from the current working directory to ~/plots by typing

```
% mv [A-Z]*.ps ~/plots/                                    ↵
```

The command

```
% rm [A-Z]*.ps                                             ↵
```

can be used to remove all files whose names begin with a capital letter. Submitting

```
% mv *.f90 ../f90Codes/                                    ↵
```

at the Unix/Linux shell prompt moves all files with extension .f90 to directory f90Codes, where the double period refers to the parent directory of f90Codes, i.e., the directory that contains f90Codes.

These are not the only forms of wildcards supported by csh or bash. Another useful wildcard, for instance, is the pattern list [a-z].ps which selects all postscript files whose names begin with a lowercase letter. The pattern list [^a-z], which filters out any single character not contained in the specified range, could be used to list only those files and directories that begin with a capital letter or a number, and the command

```
% [^b-z,A-Z]*                                              ↵
```

will list all files and directories whose names begin with an 'a'. Nothing else would be shown. The command

```
% ls ?????                                                 ↵
a.pdf
```

lists all regular (not hidden) files and directories whose names contain precisely five characters, such as for a.pdf. Finally, we mention that the command

```
% mv *.f *.for
```

will not rename all files ending with .f to files with the same prefixes, but ending in .for, as is the case for some other OSs. This is easily understood by noting that file expansion occurs before the final argument list is passed along to the mv command.

If there are no `.for` files in the working directory, `*.for` will expand to nothing and the shell command will be identical to

$$\% \text{ mv } *.f,$$

which is something very different from what was intended.

### 1.5.4 Using the C-shell history and event mechanisms

The C-shell maintains a numbered history of previously entered command lines. Because each line may consist of more than one distinct command (separated by a semicolon), the lines are called events rather simply commands. To view the shell history, type

```
% history                                    ↵
```

after entering a few commands at the shell prompt. Although bash, which we assume you are using, allows you to work back through the command history using the up-arrow and down-arrow keys, the following event designators for recalling and modifying events are still useful, in particular if the event number is part of the shell prompt, as is the case for the initial set-up on many Linux machines. The command

```
% !!                                         ↵
```

causes the shell to repeat the previous command line, while

```
% !22                                        ↵
```

will repeat the command with line number 22. Unix/Linux users often refer to an exclamation point ('!') as 'bang'. To repeat the most recently issued command line which started with an 'a', type

```
% !a                                         ↵
```

An initial sub-string of length greater than one can be used for more specificity. The command

```
% !?b                                                    ↵
```

is used to repeat the most recently issued command line which contains 'b'. Any string of characters can be used after the question mark.

### 1.5.5 Standard input, standard output, and standard error

Every program run from a shell automatically opens three files (data streams), which are standard input (stdin), standard output (stdout), and standard error (stderr). These files provide the primary means of communications between the programs. They exist for as long as a given process runs from a shell. The standard input file provides a way to send data to a process. As a default, standard input is read from the terminal keyboard. The standard output provides a means for the program to output data. As a default, standard output is written to the terminal display screen. The standard error is where the program reports any errors encountered during execution. By default, the standard error is also written to the terminal display. Below, we use the cat command with no arguments to illustrate how stdin and stdout work:

```
% cat                                                    ↵
something                                                ↵
something
something else                                           ↵
something else
^D
```

Here, the command cat run from a shell reads the lines marked red from stdin (i.e., the terminal window) and writes them, shown in blue, to stdout (also the terminal window). In other words, every line that is typed by the user is echoed by the command. A command, such as cat, which reads from stdin and writes to stdout is known as a filter.

### 1.5.6 Redirecting input and output

The power and flexibility of the stdin and stdout mechanism becomes apparent when input and output is redirection, which is implemented in the C-shell and the Bash shell. As the name suggests, redirection means that stdin and/or stdout are associated with targets other than the terminal display. Input redirection is accomplished using the '<' (less than) character, which is followed by the name of a file from which the input is to be read or extracted. Thus, the command line

```
% cat < input.dat                                           ↵
```

causes the contents of the file input.dat to be used as input for the cat command. If the content of input.dat is given by

```
% more input.dat                                            ↵
1
2
3
4
```

then feeding these numbers to cat leads to the following terminal display:

```
% cat < input.dat                                           ↵
1
2
3
4
```

Output redirection is accomplished by using the '>' (greater than) character, again followed by the name of a file to which the (standard) output of the command is to be written. Thus,

```
% cat > output.dat                                          ↵
```

will cause cat to read lines from the terminal window and copy them to the file output.dat. Care must be exercised when using output redirection because one of the first things which will happen in the above example is that the file output.dat will be clobbered. If the shell variable noclobber is set (strongly recommended for novices), then the output will not be allowed to be redirected to an already existing file. Thus, in the above example, if output.dat already exists, the shell would respond as follows,

```
% cat > output.dat                                          ↵
output.dat:  File exists
```

and the command would be aborted. The standard output from a command can also be appended to a file using the two-character sequence '≫' (no intervening spaces). Thus,

```
% cat ≫ existing_file.dat                          ↩
```

will append lines typed at the terminal to the end of existing_file.dat. From time to time it is convenient to be able to throw away the standard output of a command. Unix/Linux systems have a special file called /dev/null which is ideally suited for this purpose. Output redirection to this file, as shown in this example,

```
% cat input.dat > /dev/null                        ↩
%
```

causes the stdout output disappear entirely from the command line terminal. Only the shell prompt is returned on the terminal window.

### 1.5.7 Pipelines

It is then often possible to combine commands (programs) on the command line so that the standard output from one command is fed directly into the standard input of another. In this case we say that the output of the first command is piped into the input of the other. Here is an example:

```
% ls -1 | wc                                       ↩
63 588 3964
```

The -1 option tells the listing command ls to show regular files and directories, one per line. The command wc (which stands for word count) when invoked with no arguments, reads stdin until an end-of-file (EOF) is encountered and then prints three numbers: (1) the total number of lines in the input, (2) the total number of words in the input, and (3) the total number of characters in the input. For the above example, these numbers are 63, 588, and 3964, respectively. The pipe symbol '‖' tells the shell to connect the standard output of ls to the standard input of the wc command. The entire ls -1 ‖ wc construct is known as a pipeline. The first number (i.e., 63) which appears on the standard output is thus simply the number of regular files and directories in the current directory, where the listing is being created.

Pipelines can be made as long as desired, and once you know a few Unix/Linux commands and have mastered the basics of the C-shell history mechanism, you can easily accomplish some fairly sophisticated tasks by building up multistage pipelines.

A powerful Unix/Linux tool which searches for matching a regular expression against text in a file, multiple files, or a stream of input is the grep command. It searches for the pattern of text that is specified on the command line and prints output for the user. grep, which loosely stands for (g)lobal search for (r)egular (e)xpression with (p)rint, and has the following general syntax,

grep [options] regexp [file1 file2 …]

where regexp, which stands for regular expression, is a string that is used to describe several sequences of characters. Invoking grep with just a regular regexp as the only argument,

```
% grep regexp                                                      ↵
```

will read lines from stdin, usually the terminal window, and echo only those lines which contain the string regexp. If one or more file arguments are supplied along with regexp, then grep will search all those files for lines matching regexp and print the matching lines to standard output, which is usually the terminal window again. Thus,

```
% grep thesis *                                                    ↵
```

will print all the lines of all the regular files in the current working directory which contain the string thesis. Files in subdirectories residing in the working directory will not be searched, however. Recall that the '*' wildcard represents every string. So it can be used as the argument file for file causing the shell to search for thesis in all files in the current directory. A few more useful options to grep are worth mentioning. The first is -i, which tells grep to perform a case insensitive pattern matching. By default, grep is case sensitive. Thus,

```
% grep -i thesis mynotes                                           ↵
```

will print all lines of the text file mynotes which contain 'thesis', 'Thesis', 'THes', etc. Second, the -v option instructs grep to print all lines which do not match the pattern. An example of this is shown here,

```
% grep -v thesis mynotes                                           ↵
```

which will print all lines of text of mynotes which do not contain any of the symbols contained in student. Finally, the -n option tells grep to include a line number at the beginning of each line that is being printed. Thus,

```
% grep -in thes mynotes                          ↵
133:   Notes regarding my thesis:
325:   The date of the thesis defense is still unclear.
910:   The thesis committee consists of four members.
```

searches the file mynotes for the case insensitive pattern thes and prints all lines, together with line numbers in the first column, which contain the strings 'thes', 'Thes', 'tHes', etc. Note that multiple options can be specified with a single '-' sign followed by a string of option letters with no intervening blanks.

Next we show a few, slightly more complicated examples of how grep can be used to find strings of text. Note that when supplying a regular expression that contains characters such as '*', '#', '?', '[', or '!', which are special to the shell, the regular expression should be surrounded by single quotes to prevent shell interpretation of the shell characters. In fact, a user will not go wrong by always enclosing the regular expression in single (or double) quotes, as shown in this example:

```
% grep -owE '^[[:alnum:]]{7}' mynotes            ↵
```

This will search for, and print on the terminal window, all alphanumeric strings in mynotes that are exactly seven characters long. The command

```
% grep 's' mynotes | grep 't'                    ↵
```

prints all lines of mynotes which contain at least one 's' and one 't', such as lines containing 'student' or 'thus'. Note the use of the pipe symbol to redirect the stdout from the first grep to the stdin of the second grep. The command

```
% grep -v '^#' mynotes > output                  ↵
```

extracts all lines from file mynotes which do not have a '#' in the first column and writes them to a file named output. Pattern matching using regular expressions, as discussed just above, is a powerful tool. But it can be made even more powerful when combining it with certain extensions. Many of these extensions are implemented in a relative of grep, known as egrep. Details about egrep can be found in the man pages by typing man egrep.

### 1.5.8 Usage of quotes

Most shells, including the C-shell and the Bash shell, use three different types of quotes found on every standard keyboard. These are regular quotes (') also known as forward quotes, single quotes, or just quote, double quotes ("), and backward quotes (`) also referred to as just back-quotes. They serve distinct roles on Unix/Linux machines, which will be discussed here.

**Single quotes:** We have already encountered several situations where forward quotes have been used to quote variables. In essence, they inhibit shell evaluation of special characters and/or constructs. Here is an example. In a terminal session, let us assign variable a a numerical value of 100 and then prints the value of a with the echo $a command,

```
% set a=100                                          ↵
% echo $a                                            ↵
100
```

Next we assign the value of $a to the new variable b,

```
% set b=$a                                           ↵
% echo $b                                            ↵
100
```

and use echo $b to verify the value of $b. Now let us repeat the last steps but with $a put in quotes,

```
% set b='$a'                                         ↵
% echo $b                                            ↵
$a
```

which protects $a from shell evaluation. The command echo $b therefore does not return 100 but rather $a. Single quotes are commonly used to assign a shell variable a value which contains whitespace(s) or to protect command arguments which contain characters special to the shell (see the discussion of grep).

**Double quotes:** The double quotes function in much the same way as forward quotes, except that the shell looks inside them and evaluates both any references to the values of shell variables as well as anything sandwiched within back-quotes (see discussion of backward quotes below). An example is shown here:

```
% set a = 200                               ↵
% echo $a                                   ↵
% 200
% set string="The value of a is $a"         ↵
% echo $string                              ↵
The value of a is 200
% set string='The value of a is $a'         ↵
The value of a is $a
```

The first line assigns a numerical value to variable a, which is then printed on the terminal window. This is followed by the `set string` command line, which assigns a text message plus a numerical value, carried by a, to `string`. Thus, `echo $string` returns the assigned text message but with 200 substituted for $a. As shown by the last two lines, this is not the case if single quotes are used, in which case the text message as sandwiched between the single quotes is returned to the terminal.

**Backward quotes** The shell uses back-quotes to provide a powerful mechanism for capturing the standard output of a Unix/Linux command (or, more generally, a sequence of Unix/Linux commands) as a string, which can then be assigned to a shell variable or used as an argument for another command. Specifically, when the shell encounters a string enclosed in back-quotes, it attempts to evaluate the string as a Unix/Linux command, precisely as if the string had been entered at the shell prompt, and returns the standard output of the command as a string. In effect, the output of the command is substituted for the string and the enclosing back-quotes. Here are a few simple examples:

```
% date                                      ↵
Wed Jun 19 13:16:22 PDT 2024
% set current_date_and_time=`date`          ↵
% echo $current_date_and_time               ↵
Wed Jun 19 13:16:22 PDT 2024
```

The `date` command returns the current date and time to the window terminal. The `set` command is used to assign the current date and time to the variable `current_date_and_time`, which is then echoed back to the terminal.

Introduction to Computational Physics for Undergraduates
(Second Edition)

**Omair Zubairi and Fridolin Weber**

# Chapter 2

## Text editors

All computer programs are written is some sort of text editor. There are many text editors across multiple platforms. Notepad++ is very popular on Windows, while 'vi' is the most widely used text editor on the Linux/Unix operating system. Due to the ease of the Internet, text editors are widely available. Some others which can be used across multiple platforms include gedit, jedit, and atom, each with their unique niches. In this chapter, we focus on two of the most versatile and widely used text editors—'vi' or 'vim' (vi Improved) and Emacs.

## 2.1 Vi

'vi' is a widely used text editor for UNIX systems. It is often available when other editors are not. vi does not make use of a user-interface menu nor does it have online help. However, you can learn about vi by using the man command. vi is a Unix command and therefore is case sensitive. Some vi commands are issued in UPPERCASE. Be careful to make the distinction between uppercase commands and lowercase commands to issue the appropriate command. To create a new file or edit an existing one, you invoke the vi editor by keying: `vi filename`. The vi editor has two modes: *command mode* and *insert mode.* In command mode you can position the cursor or issue a vi command. The last line on the screen displays the name and size of your file. Command mode allows you to position your cursor or issue a command.

To issue a command in vi you use the ':' (colon) to precede the command. You enter the text mode by keying either an 'i' (insert) or an 'o' (open). In text mode you enter your text into a file. You use the <ESC> key (escape) to exit the text mode. There are several ways to save files and leave vi. To quit without saving the file key: ': q!'. To save the file and continuing editing key: ':w'. To save the file and quit vi key: ':wq'. After you have saved your file, you can also exit vi by keying: ':q'. Table 2.1 contains a list of some of the most commonly used `vi` editing commands.

doi:10.1088/978-0-7503-6493-5ch2

**Table 2.1.** Summary of Search and Replace Commands in vi

| Command | Description |
|---|---|
| **Window Movements** | |
| <ctrl >d | Scroll down |
| <ctrl >u | Scroll up |
| <ctrl >b | Page backward |
| <ctrl >f | Page forward |
| 1G | Go to first line |
| G | Go to last line |
| **Cursor Movements** | |
| H | Home (upper left-hand corner) |
| L | Lower left-hand corner |
| h | Back a character |
| j | Down a line |
| k | Up a line |
| ^ | Beginning of line |
| $ | End of line |
| **Input** | |
| a | Append after cursor |
| i | Insert before cursor |
| o | Open line below |
| O | Open line above |
| **Deletion** | |
| dd | Delete current line |
| x | Delete current character |
| **Undo** | |
| u | Undo last change |
| U | Undo all changes online |
| **Rearrangement** | |
| yy or Y | Yank (copy) line to general buffer |
| yw | Yank word to buffer |
| "ap | Put text from buffer a after cursor |
| p | Put general buffer after cursor |
| J | Join lines |
| **Search and Replace** | |
| /string | Search for string |
| n | Repeat search in the same direction |
| N | Repeat search in the opposite direction |
| :%s/old/new/ | Replace the first occurrence of 'old' with 'new' |
| :%s/old/new/g | Replace all occurrences of 'old' with "new' |
| :%s/old/new/gc | Replace all occurrences of 'old' with 'new', confirming each replacement |

## 2.2 Emacs

Emacs is a powerful and highly customizable text editor that caters to a wide range of editing needs, from basic text files to complex programming projects. Emacs is known for its extensive features, which include content-sensitive editing modes with syntax highlighting, and its ability to handle a variety of file types such as plain text, source code, and HTML.

One of the key strengths of Emacs lies in its extensibility and customization. Users can modify almost every aspect of the editor to fit their specific workflows and preferences. This flexibility is achieved through Emacs Lisp, the built-in scripting language, which allows users to add new functionalities or change existing ones.

In Emacs, commands are typically composed of a combination of keystrokes, where 'C-' denotes the Control key and 'M-' denotes the Meta key (often mapped to the Alt or Escape key). Table 2.2 contains a list of some of the most commonly used Emacs editing commands.

**Table 2.2.** List of commonly used Emacs commands.

| Command | Description |
| --- | --- |
| **Basics** | |
| C-x C-f | Find file, i.e., open/create a file in buffer |
| C-x C-s | Save the file |
| C-x C-w | Write the text to an alternate name |
| C-x C-v | Find alternate file |
| C-x i | Insert file at cursor position |
| C-x b | Create/switch buffers |
| C-x C-b | Show buffer list |
| C-x k | Kill buffer |
| C-z | Suspend emacs |
| C-x C-c | Close down Emacs |
| **Basic movement** | |
| C-f | Forward char |
| C-b | Backward char |
| C-p | Previous line |
| C-n | Nemnxt line |
| M-f | Forward one word |
| M-b | Backward one word |
| C-a | Beginning of line |
| C-e | End of line |
| C-v | One page up |
| M-v | Scroll down one page |
| M-< | Beginning of text |
| M-> | End of text |

*(Continued)*

**Table 2.2.** (*Continued*)

| Command | Description |
|---|---|
| **Editing** | |
| M-n | Repeat the following command n times |
| C-u | Repeat the following command four times |
| C-u n | Repeat n times |
| C-d | Delete a char |
| M-d | Delete word |
| M-Del | Delete word backwards |
| C-k | Kill line |
| C-Space | Set beginning mark (for region marking for example) |
| C-w | Kill (delete) the marked region |
| M-w | Copy the marked region |
| C-y | Yank (paste) the copied/killed region/line |
| M-y | Yank earlier text (cycle through kill buffer) |
| C-x C-x | Exchange cursor and mark |
| C-t | Transpose two chars |
| M-t | Transpose two words |
| C-x C-t | Transpose lines |
| M-u | Make letters uppercase in word from cursor position to end |
| M-c | Make first letter in word uppercase |
| M-l | Make letters lowercase in word |
| **Important** | |
| C-g | Quit the running/entered command |
| C-x u | Undo previous action |
| M-x revert-buffer RETURN | Undo all changes since last save |
| M-x recover-file RETURN | Recover text from an autosave-file |
| M-x recover-session RETURN | Recover multiple edited files from an autosave session |
| **Search/Replace** | |
| C-s | Search forward |
| C-r | Search backward |
| C-g | Return to where search started (if still in search mode) |
| M-% | Query replace |
| M-x query-replace-regexp | Search and replace using regular expressions |
| **Window commands** | |
| C-x 2 | Split window vertically |
| C-x o | Change to other window |
| C-x 0 | Delete window |
| C-x 1 | Close all windows except the current one |

**IOP** Publishing

# Introduction to Computational Physics for Undergraduates
## (Second Edition)

**Omair Zubairi and Fridolin Weber**

# Chapter 3

## The Fortran 90 programming language

## 3.1 Introduction

Fortran [1–3], derived from FORmula TRANslation, is a powerful and widely used programming language, which was designed specifically for numerical applications. It is the most widely used programming language in the world for numerical applications. It has achieved this position partly by being on the scene earlier than any of the other major languages and partly because it seems gradually to have evolved the features which its users, especially scientists and engineers, found most useful. Successive versions have added support for structured programming and processing of character-based data (FORTRAN 77), array programming, modular programming and generic programming (Fortran 90), high performance Fortran (Fortran 95), object-oriented programming (Fortran 2003), and concurrent programming (Fortran 2008) [4]. The latest standard of Fortran is Fortran 2018 [5, 6].

Aside from Fortran, there are a number of programming languages available to implement algorithms on computers. These languages are often optimized with a particular kind of programming/application in mind. These include Java for web-enabled and networked applications, C++ for managing large complex projects, C for system programming, COBOL for commercial transactions, and LISP for artificial intelligence programming [7]. Fortran is one of the older languages and was designed specifically for numerical applications. The language is often disparaged for its lack of power and expressiveness, but its strength lies in the fact that it effectively shields the programmer from the hardware, is easy to learn and use, produces optimal machine code that allows it to run fast, and has built-in support for the most widely used data structures in numerical computing, namely multi-dimensional arrays to represent matrices and vectors. The newer variants of the language (e.g., Fortran 90, Fortran 95, Fortran 2000, and Fortran 2018) expand the language substantially and allow more elaborate code to be written easily.

doi:10.1088/978-0-7503-6493-5ch3 3-1

Moreover, there is a large body of excellent time-tested software libraries written in Fortran. These are tremendous assets to anybody who is building an application but does not want to write every single line of code from scratch.

This section gives a brief history of the language, outlines its future prospects, and summarizes its strengths and weaknesses.

### 3.1.1 Early development

Fortran was invented by a team of programmers working for IBM in the early-1950s. This group, led by John Backus, produced the first compiler—a computer program which transforms source code into a language which the computer can understand—for an IBM 704 computer in 1957 [8]. They used the name Fortran because one of their principal aims was 'FORmula TRANslation'. But Fortran was in fact one of the very first high-level languages: it came complete with control structures and facilities for input/output. Fortran became popular quite rapidly and compilers were soon produced for other IBM machines. Before long other manufacturers were forced to design Fortran compilers for their own hardware. By 1963, all of the major manufacturers had joined in and there were dozens of different Fortran compilers in existence, many of them rather more powerful than the original.

All this resulted in a chaos of incompatible dialects. Some order was restored in 1966 when an American national standard was defined for Fortran. This was the first time that a standard had ever been produced for a computer programming language. Although it was very valuable, it hardly checked the growth of the language. Quite deliberately, the Fortran 66 standard only specified a set of language features which had to be present—it did not prevent other features being added. As time went on, these extensions proliferated and the need for a further standardization exercise became apparent. This eventually resulted in Fortran 77.[9]

### 3.1.2 Standardization

One of the most important features of Fortran programs is their portability, i.e., the ease with which they can be moved from one computer system to another. Now that each generation of hardware succeeds the previous one every few years, while good software often lasts for much longer, more and more programs need to be portable. The growth in computer networks is also encouraging the development of portable programs.

The first step in achieving portability is to ensure that a standard form of programming language is acceptable everywhere. This need is now widely recognized and has resulted in the development of standards for all the major programming languages. In practice, however, many of the new standards have been ignored and standard-conforming systems for languages such as Basic and Pascal are still very rare.

Fortunately, Fortran is in much better shape: almost all current Fortran systems are designed to conform to the standard usually called Fortran 77. This was produced in 1977 by a committee of the American National Standards Institute

(ANSI) and was subsequently adopted by the International Standards Organization (ISO). The definition was published as ANSI X3.9-1978 and ISO 1539-1980. The term 'Standard Fortran' will be used in the rest of this book to refer to Fortran 77 according to this definition.

Fortran is now one of the most widely used computer languages in the world with compilers available for almost every type of computer on the market. Since Fortran 77 is quite good at handling character strings as well as numbers and also has powerful file-handling and input/output facilities, it is suitable for a much wider range of applications than before.

### 3.1.3 Fortran 90

The ISO Standard for Fortran 90 has, officially, replaced that for Fortran 77. It introduces a wealth of new features many of them already in use in other high-level languages, which will make programming easier, and facilitate the construction of portable and robust programs. Therefore, this book will only focus on Fortran 90. However, due to the ease of adding new features rather removing old ones, many scientists and engineers still write code in Fortran 77 with perhaps a few minor extensions. All existing Fortran 77 programs will automatically conform to Fortran 90 compilers, and therefore can still run on multiple operating systems. It is advisable to use the syntax for Fortran 90 instead of Fortran 77 for two main reasons:

1. Improvements that make Fortran a more flexible and powerful programming language, with the possibility of object-oriented and well-structured programming, access to more system-level processes such as memory allocation and pointers, etc. In other words, Fortran 90 looks a lot more like C++ than the Fortran 77 standard. Fortran still has some ways to go in this respect.
2. Improvements to improve numerical capabilities, in the first place by incorporating parallel (vectorized) syntax commands. This is a major improvement in terms of scientific computing and allows the writing of universal parallel codes. More has been done on this in the updated version of Fortran 90 with Fortran 95 standard.

### 3.1.4 Strengths and weaknesses

Fortran has become popular and widespread because of its unique combination of properties. Its numerical and input/output facilities are almost unrivaled while those for logic and character handling are as good as most other languages. Fortran is simple enough that you do not need to be a computer specialist to become familiar with it fairly quickly, yet it has features, such as the independent compilation of program units, which allow it to be used on very large applications. Programs written in Fortran are also more portable than those in other major languages. The efficiency of compiled code also tends to be quite high because the language is straightforward to compile and techniques for handling Fortran have reached a considerable degree of refinement. Finally, the ease with which existing procedures can be incorporated into new software makes it especially easy to develop new programs out of old ones.

It cannot be denied, however, that Fortran has more than its fair share of weaknesses and drawbacks. Many of these have existed in Fortran since it was first invented and ought to have been eliminated long ago: examples include the six-character limit on symbolic names, the fixed statement layout, and the need to use statement labels.

Fortran also has rather liberal rules and an extensive system of default values: while this reduces programming effort, it also makes it harder for the system to detect the programmer's mistakes. In many other programming languages, for example, the data type of every variable has to be declared in advance. Fortran does not insist on this but, in consequence, if you make a spelling mistake in a variable name the compiler is likely to use two variables when you only intended to use one. Such errors can be serious but are not always easy to detect.

Fortran also lacks various control and data structures which simplify programming languages with a more modern design. These limitations, and others, are all eliminated with the advent of Fortran 90.

## 3.2 Compilers

Any source code written in a programming language such as C, Fortran, or Java, needs to be compiled from another computer program written in some other language. As mentioned earlier in this chapter, a compiler is a computer program which transforms source code into a language which the computer can understand. This is usually known as 'compiling your source code.' Your source code when compiled will create another computer program, usually referred to as an 'executable.' The program user will then 'run' the executable program, which in turn will produce the desired results from the source code. Fortran has many compilers, such as

- ifort (Intel Fortran compiler)
- pgi (Portland Group Inc. compiler)
- nagfor (Numerical algorithms Group compiler)
- af90 & af95 (Absoft Fortran Compiler)
- gfortran (GNU Fortran compiler)

where each of these compilers has its own niche and is designed for various purposes. In this text, we will focus our attention to the gfortran open source compiler, which is readily available on multiple platforms and operating systems. The gfortran compiler is also one of the more restrictive compilers allowing programmers to write code which will be compiler independent.

### 3.2.1 File extensions and compiling commands

The standard file extension for any Fortran 90 program is simply '.f90.' Using a text editor such as Vim, a Fortran program can be easily created in the Linux environment via the terminal command:

```
> vim filename.f90                                               ↩
```

Once, you have created and written your Fortran source code. The next step is to compile and run your program. To compile your Fortran program, the command requires a few things:

- The type of compiler (i.e. gfortran)
- Flags (for specifically naming your executable and for optimization purposes)
- Executable name (i.e. filename.x)
- Source code (i.e. filename.f90)

The four mentioned items are all typed in one line of your Linux terminal, such as

```
> gfortran -o filename.x filename.f90                           ↩
```

The command above will create the executable filename.x, and thus to run this executable simply type in the command:

```
> ./filename.x                                                  ↩
```

The flag option '-o' in the example above is specifically for naming conventions of your executable. The file 'filename.x' is the executable you will be creating from your 'filename.f90' source code. It is important to note that the name of your executable can be anything and does not explicitly require a file extension; however, most programmers will use some sort of extension to distinguish the executable from the source code.

If your source code does not need any external libraries or you simply have only one Fortran file, you can also compile this one Fortran file via the command:

```
> gfortran filename.f90                                         ↩
```

which will result in an automatically created executable named a.out in the Linux/ Unix environment or a.exe on Windows platforms To run this executable, the same command applies as before

```
>  ./a.out                                                          ↵
```

Every single time you compile, the automatic executables' shown here will be overwritten, and thus one has to be mindful of compilations of multiple programs.

The important thing to note is that the Linux compiler supports some features (not all) that are in the 90 standard, and which are very important to good programming. In particular, it supports the Fortran 90 structure of a DO–END DO loop, the type declaration statement with the double colon :: syntax, and the standard relational operators instead of the Fortran 77 verbose substitutions.

## 3.3 Program layout

Programs are usually written by using some text editor. Readability and cohesive flow throughout a program is not only important to the programmer but is also vital for external collaborations with multiple parties. The guidelines listed below serve as good practice for writing computer programs.

- Start your code with some sort of prologue (describing your code).
- Give compiler directions (i.e. how to compile and run your code).
- Put Fortran key words and intrinsic functions in uppercase, or at least have the starting letter capitalized.
- Put variable names in lowercase.
- Use descriptive variable names.
- Use blank lines to improve readability.
- Use multiple lines to improve readability (i.e., do not use one long line of code).
- Indent your code when possible.
- Finally, have a running commentary describing different parts of your code.

As you write more computer programs, you will start to have your own unique style of coding and you may not want to follow these rules explicitly; however, the general gist of good efficient programming lies within these general guidelines listed above.

As new text editors emerge, Fortran 90 has evolved to meet these standards. A few important improvements include:

- A line may contain up to 132 characters and may contain more than one Fortran statement provided a semicolon separates each successive pair of statements.
- Blanks are significant.
- A trailing ampersand, &, indicates a statement is continued on the next line (a maximum of 39 continuation lines is allowed). Note that comments therefore

cannot be continued and a separate exclamation mark (!) must be used for each comment line.

- Statement labels consist of up to five consecutive digits preceding the command.
- Comment lines must begin with the exclamation mark, !. Also, trailing comments (after a command) are allowed and also go after !.

The general structure of a Fortran 90 program looks something like this:

```
Declare Modules

PROGRAM name

!Comments and program information

     Variable Declarations and/or External Functions

     Body of program

END PROGRAM name

Declarations of user made functions
```

Two sample Fortran 90 programs are listed below for your reference. The first program "PROGRAM hello" is a standard program which outputs to terminal "Hello World." The second program "PROGRAM math" performs some basic mathematical operations and outputs the results to the terminal.

```
PROGRAM hello

!This simple program will output "Hello World."

     !print statement (output to terminal)

     Print*, "Hello World"

END PROGRAM hello
```

```fortran
PROGRAM math
!Simple Fortran program to take in some variables and
!perform basic !mathematical operations.

!Explicitly define all variables:
Implicit None

!Variable declarations:
Real ::  x, y, a, f1, f2, f3
Integer ::  b,c

!Assign numerical values:
x = 1.1; y = 2.5; a = -5.5; b = 10; c = 3;

!Add, subtract, multiply and divide some numbers:
f1 = (x+y)/y
f2 = (a*b) + (c-a)
f3 = (x-a)/(-a*y)

!Print results to terminal:
Print*,"x=",x
Print*,"y=",y
Print*,"a=",a
Print*,"b=",b
Print*,"c=",c
Print*,"(x+y)/y=",f1
Print*,"(a*b) + (c-a)=",f2
Print*, "(x-a)/(-a*y)=",f3

STOP;
END PROGRAM math
```

## 3.4 Variable declaration

Fortran is an implicit language and all variables need to be declared. The compiler needs to know the names, types, and sizes of the variables used in the program in order to allocate memory and optimize the performance. All variables must be

declared if the IMPLICIT NONE statement is used. Otherwise, a type is implied by the first letter:
- a...h and o...z are REAL
- i,j,k,l,m,n are INTEGER

It is highly encouraged to explicitly declare all variables by using the Implicit None statement, as shown previously in the sample program math. Explicit declaration is done with the statement:

```
type [, attributes] :: variable =[initial value]
```

### 3.4.1 Naming conventions

The rules for naming conventions for all variables in Fortran 90 are as follows:
- Names can be up to 31 characters long.
- Allowed symbols are letters a... z, numerals 0, 1, 2, ···, 9 and the underscore _.
- The first character must be a letter.
- Variables are case insensitive.
- Reserved words cannot be used (i.e., EXIT, EXP, SIN).

Table 3.1 shows some example valid and invalid variable names.

### 3.4.2 Data types

In your programs, you will use a variety of different type of variables, such as integers, floating-point numbers, string of letters, etc table 3.2 summarizes supported Fortran 90 data types.

**Table 3.1.** Variable names.

| Valid Variable Names | Invalid Variable Names |
| --- | --- |
| x,Y,t,a | x*y, x+y |
| temp11 | 33y, 11temp |
| delta_x | delta y |
| bsquared | _one_two |
| really_Longname10 | sin, log, exp |

**Table 3.2.** Fortran 90 data types.

| Data Type | Descriptions |
| --- | --- |
| Integer | Whole numbers |
| Real | Floating-point numbers (~8 digits) |
| Real*8 | Floating-point numbers (~16 digits) |
| Character | Strings of characters |
| Logical | Two-valued Boolean variables |
| Complex | Complex number |

Integer constants are of the form 1234, real constants are of the form 1234.0 or 1.234E03. The range of integer values is limited, but the exact value depends on the compiler and machine used. Real*8 constants are of the form 1.234D03 (often refer to double precision), and complex numbers which are represented by an an ordered pair of Real's are of the form (3.14,−1.0E05). Characters are enclosed in quotes of the form 'ABab' or 'S'. The Character statement requires the length of the string, which is written as a couple of ways:

```
Character (Len=10) :: position

Character (20) :: acceleration
```

In this example, the variable 'position' can hold up to 10 strings of characters. Note that the keyword Len= is optional and can be omitted. Thus, the variable 'acceleration' can hold up to 20 strings of characters.

Logical constants can have only two values: .TRUE. or .FALSE. (note the opening and closing periods). Sample syntax is provided below:

```
Real*8 :: grav=6.67E-11
Real :: pi=3.1415, rho, g = 9.81

Integer :: f = 30, mu, row, col
```

Fortran 90 attempts to make code more transferable and architecture-independent by allowing the user to specify the length (in words) of the variable through the KIND attribute (note that this keyword can be omitted):

type([KIND=]kind_num) [, attributes] :: variable list

```
Real (Kind = 8) :: au = 1.5E11
Real (8) :: double_number-2.0D3
```

The problem is that a word is defined differently on different machines. On most architectures (e.g., the PC), for example, a REAL with eight words is double precision, and with four words is a single precision float. A good programmer will make one separate record (module) for each of the machines the program runs on, and make symbolic names for the kinds that will be universal, while changing the length depending on the machine's architecture.

If you are using a 'constant' throughout your program, you may want to use the `Parameter` statement to define this. Once defined, the value cannot change throughout your program.

```
Real*8 ::  G, pi
Parameter (G=6.67E-11)
Parameter (pi=acos(-1.0))
```

Generally, you do not have to necessarily define the values of certain variables at the moment of declaration. You may easily assign numerical values after you have declared the type of variable. This may be more convenient if you have many different variables or if the value of the variable will depend on some other external source. However, all declaration of variables and parameter statements must be done prior to using those variables.

## 3.5  Basic expressions

### 3.5.1  Arithmetic operators and expressions

The most important arithmetic operators are addition +, subtraction −, multiplication *, division /, and exponentiation **. For the following examples

```
A+B-C; A*(−B); A*B/C; Z**I
```

The order of evaluation is:
1. Parentheses (innermost first)
2. Exponentiation
3. Multiplication and division
4. Addition and subtraction

For example, the expression `A*B−C/D` is evaluated as if it were written as `(A*B)` `− (C/D)`. Where two operators have the same priority, the order of evaluation is left to right. For example, `A/B*C` is evaluated as `(A/B) *C`, which is not equivalent to `A/ (B*C)`, of course. It is important to notice that the result of an operation between floats is a float, between integers is an integer, and between floats and integers is a float. This means, for instance, that 1/5 gives the integer 0, while 1./5. gives 0.2 in single precision, and `1D0/5D0` gives 0.2 in double precision. In general, if I is an integer (say 4), use `I.0` (4.0) to make it a floating-point number.

### 3.5.2 Relational operators

Many type of programs require the user and the code to make choices. These choices will depend on certain conditions set by the code. If the conditions are met, certain outcomes will happen, otherwise other outcomes will happen. These logical conditions are are categorized as:

- Equal
- Greater or equal to
- Greater than
- Less than or equal to
- Less than
- Not equal to
- Not
- And
- Or
- Equivalent
- Not equivalent

Logical operators Fortran 90 syntax for these conditions are described in table 3.3. For example, (A >= B) is true if the value of the variable A is greater or equal to the value of B. Operators can be applied between mixed types, and as with arithmetic operators all the arguments are converted to the highest type (integer $\rightarrow$ real $\rightarrow$ double $\rightarrow$ complex).

### 3.5.3 Logical expressions

You can combine relational expressions and other 'true–false' valued expressions and variables together to form logical expressions. This is done using the logical operators shown in table 3.3. Since some of you may not be familiar with Boolean logic, an overview of Boolean logic is provided in table 3.4 (T stands for true, F for

**Table 3.3.** Overview of relational operators.

| Relational Operators | Meaning |
| --- | --- |
| == | Equal to |
| >= | Greater or equal to |
| > | Greater than |
| <= | Less than or equal to |
| < | Less than |
| = | Not equal to |
| .NOT. | Logical negation |
| .AND. | Logical and |
| .OR. | Logical inclusive or |
| .EQV. | Logical equivalence |
| .NEQV. | Logical non-equivalence (exclusive or) |

**Table 3.4.** Overview of boolean logic.

| x | y | .NOT.x | x .AND. y | x .OR. y | x.EQV. y | x .NEQV. y |
|---|---|--------|-----------|----------|----------|------------|
| F | F | T | F | F | T | F |
| T | F | F | F | T | F | T |
| F | T | T | F | T | F | T |
| T | T | F | T | T | T | F |

false). The only rule that you need to know (in case of doubt use excessive bracketing, if necessary) is that arithmetic operators take precedence over relational operators, which take precedence over logical operators. Thus, these two lines are equivalent; however, it is usually safer to use the second line.

```
x == y .OR. b > pos .AND. acceleration < g

(x == y) .OR. (b > pos) .AND. (acceleration < g)
```

## 3.6 Input and output

### 3.6.1 The READ statement

Most programs require that the user input some data through the keyboard or that the program prints some result on the monitor. User input is achieved through the statement READ, with structure(s)

```
READ*, {input list}

READ ([UNIT=]unit type, [FMT=]format){ input list}
```

The unit type is a number (e.g., a 5 for keyboard) and the format type is * for a format–free input. For example,

```
READ (UNIT=5, FMT=*) a, b, c
```

expects the user to input three numbers separated with commas.

### 3.6.2 The WRITE statement

Write to the screen (or printer) is done using the command WRITE, which can take the form:

```
PRINT *, {output list}

WRITE ([UNIT=unit type], [FMT=]format) { output list}
```

The unit type is again a number, usually 6 for the monitor. For example,

```
WRITE (6,*) 'The total is:', total
```

prints the message in the quotes and then prints the value of the value total.

### 3.6.3 The FORMAT specification

The format can be specified by the user in two different ways. First, the format can be a number giving the label (in columns 1–5 of the program) of the line in the program containing the call:

```
FORMAT(format sequence)
```

or the format can be a string containing the format sequence, with the syntax

```
FMT=('format sequence')
```

The format sequence is a beast of its own. It is a list of data descriptors and control functions, which specify what type of data is being printed is and how it should be printed, and specify things such as new lines.

**Data types and format specifiers**

Table 3.5 gives the format descriptors for various types of data. The capital letters in table 3.5 have specific meanings and are described below:

- I – integers
- F – a fixed-point floating number
- E – a float in scientific (exponential) notation
- G – either F or G, depending on the magnitude of the number

**Table 3.5.** Format specifiers.

| Data Type | Declaration Syntax | Format Descriptors |
|---|---|---|
| Integer | INTEGER | I$w$, I$w$. $m$ |
| Floating point (Real) | REAL | E$w$. $d$, E$w$. $d$ E$e$, F$w$. $d$, G$w$. $d$, G$w$. $d$ E$e$ |
| Double precision | DOUBLE PRECISION | D$w$. $d$ |
| Logical | LOGICAL | L$w$ |
| Character | CHARACTER(LEN=length) | A, A$w$ |

- D – double-precision scientific notation
- L – logical
- A – character

with the lowercase letters meaning:
- $w$ – the total field width
- $m$ – the minimum number of digits
- $d$ – number of digits after the decimal point (precision)
- $e$ – number of digits in the exponent

For example,

```
WRITE(UNIT=*, FMT=10) 'The frequency is', f, ' Hz'
10 FORMAT(1X, A, F10.5, A)
```

displays the value of the frequency variable $f$ with a maximum of 15 digits, five decimal digits, an indent of 1 column in the beginning, and a suitable string message.

### 3.6.4 File input and output (I/O)

One of the areas in which Fortran 90 (besides number crunching) is better than most other programming languages is file input and output. In most programming languages, file does not refer just to hard disk files but to any peripheral device that one can read from or write data to (e.g., keyboard, monitor, printer, disks). Therefore, file I/O is still done using the READ and WRITE commands (usually), but with a UNIT number that specifies the device or the file the data is being read from or written to.

**OPEN**

To assign specific UNIT numbers to certain files and open (or create) the file, use the OPEN statement:

```
OPEN(UNIT=Integer, FILE='filename', STATUS=literal)
```

The literal keyword can be 'NEW' for new files, 'OLD' for existing files, 'UNKNOWN' if you are not sure, or 'SCRATCH' for a temporary files. There is one more important keyword mentioning here. The keyword ACCESS=access which would go after the STATUS keyword is one of 'SEQUENTIAL' or 'DIRECT'. The distinction is very important, but beginners should use the default of sequential files, which are simply files in which the data are written and read from line-by-line.

**CLOSE**

After we are done using the file, it is highly recommended that the file be closed using:

```
CLOSE(UNIT=number, STATUS=status, ...)
```

where the unit is the number (must be an integer value) of the file, and the default is 'KEEP' to save the file or 'DELETE' to delete it (the former is the default).

## 3.7 Control structures

Control structures determine the program flow, i.e., the sequence of execution of the commands (other than the default ordering in the program file). In Fortran 90 these are the IF blocks and the DO loops.

### 3.7.1 IF-blocks

An IF-block is a multi-branched control structure which takes the execution to different branches depending on the value of one or several condition statements. The general structure of IF blocks is as follows:

```
IF(First condition statement) THEN

    First sequence of commands

ELSE IF (Second condition statement) THEN

    Second sequence of commands

ELSE IF ...
...
ELSE

    Alternative sequence of commands
END IF
```

Any but the first IF can be omitted. The $n$th (ELSE) IF sequence of statements is evaluated if its condition statement is true and if none of the preceding conditions were true. In other words, once a condition statement is found true, its sequence of commands is evaluated and the IF-block is exited. The ELSE sequence of commands is evaluated if none of the condition statements are true. For example, to find the sign of a number (signum$(x) = 1$ for $x > 0$, $-1$ for $x < 0$, and 0 for $x = 0$) we can use the following construction:

```
IF (number < 0) THEN

     signum=-1

ELSE IF (number > 0) THEN

     signum=1

ELSE

     signum=0

END IF
```

Many IF statements are one liners, and more clarity and less typing are achieved using an abbreviated IF one-liner:

```
IF (Condition statement) Statement to be executed
```

which simply omits the THEN and END IF in the usual conditional IF-block. The following is a sample program which illustrates the If–Then–Else structure:

```
Program TryIf
write(*,100)
Read(*,*) X
If (X < 0) Then
    write(*,*) X, 'is a negative number'
elseif (X > 0) Then
    write(*,*) X, 'is a positive number'
else
        write(*,*) 'X is 0'
    endif
100  format('Please input a number: ',$)
        end program TryIf
```

### 3.7.2 DO loops

Loops are blocks of commands to be repeated, as illustrated below:

```
Do i = start, finish, increment

    Executable Statements

End Do
```

where $i$ must be an integer value and is called the control variable. The increment keyword is optional. If it is specifically not listed, the default increment is 1. For example,

```
Do i = 1, 10

    Print*, i**2

End Do
```

the loop variable $i$ will start at i=1 and print 1*1=1, then the loop variable will go to the next value determined by the increment (in this case, the increment is 1), which is i=2 and print 2*2=4, and thus will continue to the value of 10.

Instead of incrementing values of 1, what if the increment is some other non-integer value, such as 0.5? This requires a little bit of thought. Having this increment be implemented correctly in the loop control structure is a very important part of iterating correctly. There are a few important concepts to think about here:

1. First, If we are going to increment in units of '0.5' we are going to have to somehow declare that, this is commonly known as a 'step-size.'
2. Second, since we are going to be incrementing half a unit, we are going to need to double our iterations.
3. Finally, we are going to need some expression that will allow us to print out the correct results.

The sample code below illustrates these three important concepts:

```
Program loops
        Implicit None

        Real :: dn, f
        Integer ::  n

        dn=0.5

        Do n=1,20
            f=dn*n
            Print*, f**2
        End Do
        Stop
End Program loops
```

In the sample code above, dn is our step-size and the variable $f$ is our function which is defined by our step-size times the loop variable $n$. Our last entry ($f_{20} = (0.5)(20) = 10 = 10*10 = 100$) is just as before. All we have changed is the increment and thus the 'amount' of data generated.

You may be wondering, why did we apply these certain extra steps? Could we not simply just add our increment, such as:

```
Do i = 1, 10, 0.5

    Executable Statements

End Do
```

and obtain the same results with fewer lines of code? The problem with the above example program is that that it will only work for certain compilers which will not distinguish between Real's and Integer's as loop control statements. As mentioned earlier in this chapter, good and efficient programming should always be compiler independent, i.e., the sample full program shown previously in which we explicitly declared our 'step-size will compile using any compiler, while if we used Do i = 1, 10, 0.5 instead it will only work on certain compilers.

**The** Do While **loop**

You can loop over quantities with a condition as well by using the Do While statement:

```
Do While (Condition)

    Executable Statements

End Do
```

This loop will execute as long as the condition stays true. A sample program is provided for your reference:

```
Program dowhile
        Implicit None

        Real :: none
        Integer :: counter

        counter = 0

        Do While (counter < 10)
            counter = counter + 1
            Print*, counter
        End Do
Stop; End Program dowhile
```

In this sample program, we are iterating the variable counter as long as the counter variable is less than 10. In the first iteration, the counter is zero, and thus the variable then equals counter $= 0 + 1 = 1$. The second iteration would be counter $= 1 + 1 = 2$, and so on and so forth as long as the condition of counter $< 10$. Once the the counter reaches a value which is equal or greater than 10, the program will stop and end.

### 3.7.3 Nested loops

Just as single, loops as in the previous section, control structure with loops can also be nested. The standard statements for nested loops are as follows:

```
Do i = 1, 10

    Do j = 1, 10

    Executable Statements

    End Do

End Do
```

Nested loops start 'inner' and go 'outer.' For example,

```
Do i = 1, 10

    Do j = 1, 10

    Print*, i*j

    End Do

End Do
```

the program is iterating the loop variables $j$ and then $i$. More specifically, our iterations will be as described by the expressions in equations (3.1) and (3.2).

$$
\begin{aligned}
i = 1, \; j = 1 &\;\Rightarrow\; i \cdot j = 1 \\
i = 1, \; j = 2 &\;\Rightarrow\; i \cdot j = 2 \\
i = 1, \; j = 3 &\;\Rightarrow\; i \cdot j = 3 \\
\vdots \quad \vdots \quad \vdots \quad & \vdots \quad \vdots \quad \vdots \quad \vdots \\
i = 1, \; j = 10 &\;\Rightarrow\; i \cdot j = 10
\end{aligned}
\tag{3.1}
$$

$$
\begin{aligned}
i = 2, \; j = 1 &\;\Rightarrow\; i \cdot j = 1 \\
i = 2, \; j = 2 &\;\Rightarrow\; i \cdot j = 2 \\
i = 2, \; j = 3 &\;\Rightarrow\; i \cdot j = 6 \\
\vdots \quad \vdots \quad \vdots \quad & \vdots \quad \vdots \quad \vdots \quad \vdots \\
i = 2, \; j = 10 &\;\Rightarrow\; i \cdot j = 20
\end{aligned}
\tag{3.2}
$$

## 3.8 Modular programming

### 3.8.1 Intrinsic functions

There is a library of intrinsic functions available to any Fortran program. These functions are invoked by using the function name followed by its parenthesized list of parameters:

```
function name({list of parameters})
```

A function is said to return a value based on the values passed to it through the arguments. Some of the more important intrinsic functions are ABS, ACOS, COS, DOT_PRODUCT, EXP, INT, LEN, LOG, SIN, and SQRT. See appendix B for a fuller list and details. Many intrinsic functions are generic in the sense that the functions may be used with different (but not mixed) data types as parameters to the function (e.g., ABS, COS, MAX, MIN). Some functions, however, such as the float-valued functions SQRT, SIN, and LOG, accept only certain types of arguments. For instance, SQRT(4) is invalid; one needs to type SQRT(4.0) instead. Some functions have different versions for different types of arguments (see appendix A), and the generic function calls these specialized functions depending on the types of the arguments. For example, to compute the square root of a given integer or any floating-point number (even complex), we simply write root=SQRT (1.0*number).

### 3.8.2 Intrinsic subroutines

Subroutines are very similar to functions, but there is an important distinction. Functions return one value and are not recommended to change the values of their parameters. Subroutines do not return a value explicitly, but execute a well-defined group of statements (activity) and can freely change the values of their parameters. They should also be used when we wish to return more than one value. Subroutines are invoked using a CALL statement,

```
CALL name({list of arguments})
```

Some of the intrinsic subroutines include DATE_AND_TIME, MVBITS, RANDOM_NUMBER, and RANDOM_SEED, SYSTEM_CLOCK.

### 3.8.3 External functions

You can define your own functions, usually after the END of the program. The layout for functions is

```
type FUNCTION name({dummy arguments})
     local variable declaration
     name = expression
     body of function continued if need ...
END FUNCTION name
```

The type of the name identifies the type of the result that will be returned by the function. The result of the function is the value of the function that is assigned to the name of the function within the body of the function. Therefore, the function must contain at least one assignment of a value to the name of the function. The dummy arguments are a list of constants, variables, and even procedures that are accessible from within the body of the function. When the function is used, it will have corresponding actual arguments that must be of the same type and length as the dummy arguments, but not necessarily the same names. No type checking is done during compilation or run time, and you will get some very weird errors if the types do not match. All arguments are passed using 'call by reference' (like VAR arg in Pascal or &arg in C++). Changing the value of an argument in the subroutine changes the value of the corresponding variable in the calling program. For example, to define a function that computes the gravitational force between two bodies, we define a function NEWTON as follows:

```
REAL FUNCTION Newton(m1, m2, r)
     REAL (Kind=8) ::  gamma=6.672E(-11), r
     REAL ::  m1, m2
     Newton = -gamma*m1*m1/r**2
END Newton
```

A sample code illustrating the use of external functions is provided for your reference. In this program, the code reads in three real numbers and calculates the average. The user defined function ave returns a value to VAL.

```
PROGRAM Average

    IMPLICIT NONE
    REAL :: TEST1, TEST2, TEST3, ave, VAL
    PRINT*, 'Enter three numbers, separated by a comma'
    READ *, TEST1, TEST2, TEST3
    VAL = ave(TEST1,TEST2,TEST3)
    PRINT *, VAL
    STOP
    END PROGRAM Average

FUNCTION ave(X,Y,Z)
    IMPLICIT NONE
    REAL :: X, Y, Z, ave

    ave = (X+Y+Z) / 3.0
RETURN
END FUNCTION ave
```

### 3.8.4 External subroutines

The structure of a subroutine is

```
SUBROUTINE name({dummy arguments})
    local variable declaration
    body of subroutine ...
END SUBROUTINE name
```

Subroutines are accessed by using the CALL statement, and are in most respects similar to functions. When a subroutine is called, the dummy arguments in the subroutine become alias names for the actual arguments in the calling statement, i.e., they represent the same physical location in memory. Thus, if the dummy arguments are modified within the subroutine, then so are the actual arguments in the calling statement.

The sample program provided earlier illustrating functions can be easily modified to include an external subroutine. See below for for the full program, which includes an external subroutine.

```
PROGRAM Average

    IMPLICIT NONE
    REAL :: TEST1, TEST2, TEST3, VAL
    PRINT*, 'Enter three numbers, separated by a comma'
    READ *, TEST1, TEST2, TEST3
    CALL AVG(TEST1,TEST2,TEST3, VAL)
    PRINT *, VAL
    STOP
    END PROGRAM Average

SUBROUTINE AVG(X,Y,Z,V)
    IMPLICIT NONE
    REAL :: X, Y, Z, V

    V = (X+Y+Z) / 3.0
RETURN
END SUBROUTINE AVG
```

All well-defined functions should have a single-point entry and a single-point exit, but in some situations there may be a need to terminate a procedure (e.g. in case of error). This is done by simply calling RETURN, which stops the execution of the function and returns control to the calling routine.

After the function is finished, the values of the local variables are lost. If they are to be used in a later call of the function, this should be made clear by saving the values of these between function calls (static allocation):

```
SAVE {of values to be saved}
```

When a function is passed on as an argument to another function, and in some circumstances when we wish to link a named block data subprogram into the final executable, the function has to be declared as either intrinsic or external, via the following statements:

```
INTERNAL {list of function names}
EXTERNAL {list of function names}
```

### 3.8.5 Program units

Each program consists of program units, called *scoping units*. These can be the main program, subprograms, procedures, etc. The important thing to notice is that each scoping unit is independent in terms of its variable space, so that it is good to try to put well-defined and more or less isolated groups of statements into separates units. This way, you can invoke that unit from various programs, without worrying about possible variable conflicts.

Any unit that is placed (defined) within another unit is a *sub-unit* and has access to all the variables in the unit it is a part of. Therefore, procedures that interact with the main program variables should be placed inside the PROGRAM using the CONTAINS command. These are *internal procedures*, as contrasted to *external procedures*, which are defined outside of the PROGRAM unit and have their own variable workspace. The communication, or *interface*, between the main program or subprogram and the external procedure is done through the argument list. Data is copied from actual arguments to dummy arguments upon the invocation of the procedure (this is not always true), and from dummy argument to actual argument upon exiting the procedure.

Another way of sharing of data between procedures is the usage of modules, which are new units introduced in Fortran 90, facilitating global variables and procedures. We now describe in more detail some of the new features of the Fortran 90 standard.

### 3.8.6 Internal procedures

Internal procedures should be placed after the CONTAINS command. They can be a part of any program unit, like the main program, a subprogram or a procedure (usually no more than 2–3 levels of nesting). For example, here is how to make a procedure that quickly zeros all the integer counters that we use in the main program:

```
PROGRAM counters
    INTEGER :: i, j, k, l, m, n
    CALL Zero_counters ! Zero all the counters
    CONTAINS

SUBROUTINE Zero_counters
    i=0; j=0; k=0; l=0; m=0; n=0;
END SUBROUTINE Zero_counters

END PROGRAM counters
```

### 3.8.7 External procedures

As already explained, external procedures are usually placed outside the main program (in a separate file, a module, the same file, etc) When the compiler compiles the program, it can detect a great deal more errors if it is told what the procedure's interface (argument list) is. This is done with the INTERFACE block, which contains the declarations of the external functions (it can be placed inside the main program or in a separate module). Also, the compiler can be told what the intention of the arguments of a subroutine is by using the INTENT data attribute with one of the following arguments:

- IN for arguments that should not be altered in the subroutine, but only pass data to the procedure.
- OUT for arguments that need to be assigned values in the subroutine.
- INOUT for arguments that pass data both in and out of the procedures.

For example, here is an interface to a subroutine that returns the time in seconds given the time in hours, minutes, and seconds:

```
INTERFACE
SUBROUTINE convert_time(hour,minute,second,time)
    INTEGER, INTENT(IN)::  hour, minute, second
    INTEGER, INTENT(OUT)::  time
END SUBROUTINE convert_time
END INTERFACE
```

### 3.8.8 Modules

Modules are a new program unit in Fortran 90. They are mostly used to enclose data (global variables) and INTERFACE declarations of functions that several other program units need to share. One can also put full procedure bodies inside a module (so-called module programs). The modules are defined as

```
MODULE name
    Global variable declarations
    INTERFACE blocks
    ...
    ...
END MODULE name
```

If we want a program unit to have access to the variables and procedures defined in the module, we use the USE command. If we have lots of procedure interfaces in the module (for example, a subroutine library) but would want to use only a few of these procedures, we specify this with the ONLY attribute (this is also a way to make some global variables not accessible to all program units and is good programming practice):

```
USE module name [, ONLY: {procedure names}]
```

For example, we can store the total number of floating-point operations (flops) in a program in a module,

```
MODULE flops
    INTEGER(KIND=8) ::  flops_count
END MODULE flops
```

and then update this number from any program unit that contains the statement

```
USE flops
```

## 3.9 Arrays

Arrays are collections of data of the same type. They are the most important data type in scientific computing where we usually deal with a great number of data (points).

### 3.9.1 Declaration of arrays

Arrays are declared in the same way as ordinary variables, with lower and upper bounds along with important attributes

```
type [,attributes] array({Lower bound:Upper bound})
```

These attributes are

- DIMENSION (Lower bound:Upper bound), which enables the dimension to be specified, and the lower and upper count can now be actual workspace variables, thus enabling the existence of *automatic arrays* in procedures, which come to exist with different dimensions each time the procedure is called. Also, it is very important to note that the dimensions do not have to be specified at compile time, in which case they are substituted with the ellipse symbol. If the dimensions of an array argument to a procedure are not specified, the procedure must have an INTERFACE block in the calling unit to ensure that the compiler knows that the array has to be passed along with its dimensions.
- ALLOCATABLE, which identifies the array as allocatable.
- TARGET, which identifies the array as a target for pointers.

For example, to declare an array used to store your income for several years and all the months in each year, one can use

```
REAL, DIMENSION(12,1995:1999) ::  income
```

### 3.9.2 Vectors

Vectors are one-dimensional arrays which are mainly used to access sub-arrays of a given array (see the next sub-section). It differs from an array only in the way it is assigned a value by using constructors:

```
vector = (/{lower:upper:step],.../})
```

Where list may be a list of values of the appropriate type, namely, a variable expression, array expression, implied DO loop, or any combination of these. Here is an example of a vector that has the value *vector* = (1, 3, 5, 7, 9, 6, 2, 4 ,, 8, 16):

```
INTEGER ::  vector(6)=(/1:9:2,6,(2**i,i=1,4)/)
```

### 3.9.3 Using arrays

Arrays in Fortran 90 can be accessed on an element-by-element basis, in addition to access of whole sections by the following construct:

```
array({start:end:step})

array(vector)
```

Any of the starting, ending, and step values for the indices can be omitted so long as there is one ellipse. Examples are illustrated in the following:

```
REAL ::  array(50,50), vector(3)=(/1 7 37/)

!This accesses the section of the array from rows 1-20
!and columns 5-10, returning an array of size [20,5]
!WRITE(*,*) array(1:20,5:10)

!This accesses all the even-numbered rows
WRITE(*,*) array(2:50:2,:)

!And this accesses the elements at the intersection of
!rows 1, 7 and 37 with columns 1, 7 and 37
WRITE(*,*) array(vector,vector)
```

### 3.9.4 Array operations

A very important new feature in Fortran 90 is the idea of array-array operations between *conformable arrays*. Two arrays are said to be conformable if they have the same size (not necessarily the same row or column numbering), or if one of the two arrays is a scalar. An operation between two conformed arrays is carried between the elements of the two arrays individually. For example,

```
REAL, DIMENSION(20,20) ::  A, B, C
C=A+B
```

assigns to each element in C the value of the sum of the corresponding elements of A and B. Thus, you may say that this is equivalent to a DO loop like

```
DO i=1,20
   DO j=1,20
      C(i,j)=A(i,j)+B(i,j)
   END DO
END DO
```

But this is not quite true, and it is important to understand the difference. The matrix statement C=A+B does not span in time—it spans in space. In other words, as far as summing two matrices goes, the order in which the summation is done is arbitrary. On the other hand, the double DO-loop structure spans in time, not in space. Therefore, the double loop is an unnatural translation of the matrix statement. The reason that nested looping is the traditional way of performing matrix operations is that until recently most computer people used so-called von Neuman sequential machines, in which there is a single processor that does things in a time-ordered fashion. However, today the concept of parallel, or multi-processor, machines is an essential one, so that nested do loops should be avoided whenever an equivalent array-array (vectorized) statement exists.

When the compiler sees a statement C=A+B for matrices, it automatically optimizes the process for the specific machine using the fact that the operation is not sequential. How it does this is fortunately not our concern. It may use several processors to do this if the machine has more than one, it may access the elements column or tow-wise, use pointers (those that are more experienced may know that one of the fastest operations is the incrementation of a pointer, not an integer, by one, so that an array can be accessed most quickly with a sweeping pointer), etc. The lesson from all this is that you should always try to formulate your code in vectorized statements.

### 3.9.5 Elemental functions

To continue the previous discussion, let us assume we need to take the sine of all the elements of the array A. Again, this is not a time-ordered operation, and Fortran 90 offers a very effective way of doing this. We can simply use SIN(A) to perform this operation, because SIN, like most other numerical functions in Fortran 90, is an elemental procedure that acts on each element of an array. Thus, a statement such as

```
C(5:10,:)  = SIN(A(2:10:2,:))  + 2.0 * B(5:10,:)
```

performs the very complicated matrix operation of assigning to the 5-10$^{th}$ rows of C the sum of the sine of the elements of A in the even rows of A up to the 10$^{th}$ row and the doubled value of the elements of the matrix B in columns 5-10.

### 3.9.6 The WHERE statement

The WHERE construct is a very useful control structure for performing an operation only on certain elements of an array that satisfy a given condition. The structure is as follows:

```
WHERE (array condition)
    Body of structure
END WHERE
```

For example, the following operation

```
DO j = 1,3
   DO i = 1,4
      IF(array(i,j) > 0) array(i,j) = LOG(array(i,j))
   END DO
END DO
```

can be done much more cleanly using the WHERE construct:

```
WHERE (array > 0) array = LOG(array)
```

Another example is

```
REAL, DIMENSION(20,20) ::  A, B, C
WHERE (B>0.0)
   C=A/B
END WHERE
```

which assigns each element of the matrix C the quotient of the corresponding elements of matrices A and B, so long as the element of B is nonzero. It should be noted again that the WHERE construct is not equivalent to a DO-loop with a nested IF statement because a WHERE loop is not time-ordered and can be performed in parallel. A major limitation of Fortran 90 is that WHERE constructs cannot be nested. This is corrected in Fortran 95, where WHERE and FORALL structures can be nested at any level.

The last example of a full program shown below assigns given values to a matrix array. The values are either 1 or −9.999 999, depending on whether eval has the value 'true' of 'false':

```
PROGRAM WHERE_Example

IMPLICIT NONE
REAL, DIMENSION(2,3) :: array = (/0, -1, 2, -3, 4, -5/)
LOGICAL, DIMENSION(2,3)::eval=(/.true.,.false.,.true., &
.true.,.false.,.false./)

WHERE (eval)
array = 1
ELSEWHERE
array = -9.999999
END WHERE

WRITE (*,'(3F15.6)') array

END PROGRAM WHERE_Example
```

The result is
```
1.000000-9.999999 1.000 000
1.000 000-9.000000-9.999999
```

### 3.9.7 FORALL (Fortran 95)

The FORALL structure is essentially a more general version of the WHERE construct. It may also be considered as a space-spanning equivalent to nested DO-loops that perform complicated operations whose time execution must be arbitrary. The format of the FORALL statement is

```
FORALL (index=lower bound:upper bound:increment)
statement
```

Two full sample programs are shown below for your reference:

```
program where_construct
implicit none

integer, dimension(5) :: a = (/ 1, 2, 3, 4, 5 /)
integer :: i

forall(i = 1:4, a(i) > 2) a(i) = 0

write(*, *) a
end program where_construct
```

The result is 1 2 0 0 5

```
program where_construct
implicit none

integer, dimension(5) :: a = (/ 1, 2, 3, 4, 5 /)
integer :: i

forall(i = 1:5:2) a(i) = -1.

write(*, *) a
end program where_construct
```

The result is-1 2-1 4-1

### 3.9.8 Array intrinsic functions

Fortran 90 contains a lot of new intrinsic functions for arrays. It is beyond the scope of this text to explain all them in detail. However, here we give a brief list of some of them and what they do, so you know what to look for when you need them:

- SIZE (array, [dimension]) is probably the most important inquiry function for array which finds the rank of an array along the specified dimension (or total number of elements). Use SHAPE(array) to get the shape of the array as an array to integer ranks.

- SUM (array, [dimension]) and PRODUCT(array, [dimension]) return the sum or product of the elements of an array along the specified dimension.
- MINVAL (array, [dimension]) and MAXVAL(array, [dimension]) give the minimum and maximum value of an array along the specified dimension.
- MATMUL (first array, second array) is a very important function that returns the matrix product of two arrays (optimized for parallelization when possible).
- DOT_PRODUCT (first vector, second vector) finds the dot product of two vectors.
- TRANSPOSE (matrix) gives the transpose of a two-dimensional matrix.

### 3.9.9 Allocatable arrays

An essential improvement in Fortran 90 is memory management and the introduction of allocatable arrays, specified with the ALLOCATABLE attribute. These arrays do not have a specified dimension and get allocated or deallocated using the following commands

```
ALLOCATE(array({dimensions}))
DEALLOCATE(array)
```

For example,

```
REAL, DIMENSION(:,:), ALLOCATABLE ::  array
ALLOCATE(array(10,40))
DEALLOCATE(array)
```

first reserves memory for a 10 by 40 array, and then frees the memory to the memory pool.

A full sample program illustrating dynamic memory allocation is provided below. In this program, you can create a one-dimensional vector of size dim_size. The program user then will fill in the particular entries of the vector and the program will print out the size of the vector and its entries.

```
Program Testalloc
        Implicit None

        Integer :: i, dim_size
        Integer, Allocatable,Dimension(:):: vec


        Print *,'Enter the number of elements in the vector:'
        Read *,dim_size

        Allocate(vec(dim_size))

        Print *,'The size of your vector is:',dim_size
        Print*,"*****"
        Print*,'Enter each entry of your vector:'

        Do i=1,dim_size
           Read *,vec(i)
        End Do

        Print *,'This is your vector'

        Do i=1,dim_size
           Print *,vec(i)
        End do

        Deallocate(vec)

        End Program Testalloc
```

### 3.9.10 Pointers

A pointer can be understood as a variable that points to a memory location (or locations). In particular, a pointer can point to a block of memory that we want to use to store data, or, more importantly, it can point to the memory location of another variable, called the target of the pointer. A pointer is assigned a target via

```
pointer=>target variable of array
```

When a pointer is declared, it is born in an undefined status, when it is assigned a target it becomes associated, and to avoid inadvertent misuse we can make it disassociated with

```
NULLIFY(pointer)
```

For example,

```
REAL, DIMENSION(:), POINTER ::  one_row
REAL, DIMENSION(50,50), TARGET ::  array
one_row=>array(5,:)
NULLIFY(one_row)
```

associates the pointer to the fifth row of the array. This is a useful construction because it is more efficient and easy to manipulate than, let us say, keeping in memory the number 5 to know which row we want to reference.

## References

[1] Etter D M 1995 *Fortran 90 for Engineers* 1st edn (New York: Wiley)
[2] Chapman S J 2007 *Fortran 95/2003 for Scientists and Engineers* 3rd edn (New York: McGraw-Hill)
[3] Chapman S J 2003 *Fortran 90/95 for Scientists and Engineers* 2nd edn (New York: McGraw-Hill)
[4] Wikipedia contributors Fortran https://en.wikipedia.org/wiki/Fortran (accessed 23 June 2024)
[5] ISO/IEC JTC 1/SC 22/WG 5 2018 *Information Technology–Programming Languages–Fortran* 5th edn
[6] Metcalf M, Reid J and Cohen M 2018 *Modern Fortran Explained* 4th edn (New York: Oxford University Press)
[7] Moreira J E, Midkiff S P and Gupta M 1998 A comparison of Java, C/C++, and FORTRAN for numerical computing *IEEE Antennas and Propagation Magazine* **40** 102–5
[8] Backus C 2019 The syntax and semantics of the proposed international algebraic language of the Zurich ACM-GAMM Conference *Proc. Int. Conf. Inf. Process.* **58** 125–32 Also known as the Fortran I language specification
[9] Page C G 2005 *Professional Programmer's Guide to Fortran 77* University of Leicester, US, http://www-mdp.eng.cam.ac.uk/web/CD/engapps/programming/fortran/

**IOP** Publishing

# Introduction to Computational Physics for Undergraduates (Second Edition)

**Omair Zubairi and Fridolin Weber**

# Chapter 4

## Numerical techniques

In the rapidly advancing fields of science, technology, engineering, and mathematics (STEM), computational physics plays an essential role in solving complex problems that are often impossible to address analytically. The power of computational techniques lies in their ability to approximate solutions to equations that describe a wide range of physical phenomena. From the motion of planets to the behavior of quantum particles, these equations often defy exact solutions, necessitating the use of numerical methods. In this chapter, you will be introduced to several numerical techniques that are widely used to solve equations numerically.

## 4.1 Curve fitting—method of least squares

The method of least squares assumes that the best-fit curve of a given type is the curve that minimizes the sum of the deviations squared from a given set of data. Let us suppose that the data points are $(x_1, y_1)$, $(x_2, y_2)$, $\cdots$, $(x_n, y_n)$, where $x_i$ is the independent variable and $y_i$ is the dependent variable and $i = 1, \ldots, N$, where $N$ denotes the number of data points. The fitting curve $f(x)$ has the deviation (error) $d_i$ from each data point $y_i$, that is, $d_1 = y_1 - f(x_1)$, $d_2 = y_2 - f(x_2)$, $\cdots$, $d_N = y_N - f(x_N)$. According to the method of least squares, the best fitting curve is determined by the condition that $\Pi \equiv d_1^2 + d_2^2 + \ldots + d_N^2$ is a minimum. Mathematically, this can be expressed as

$$\Pi \equiv \sum_{i=1}^{N} d_i^2 = \sum_{i=1}^{N} (y_i - f(x_i))^2 = \text{minimum}, \qquad (4.1)$$

which leads to a system of coupled linear equations for the unknown coefficients contained in the ansatz for $f(x)$. The simplest choices for $f(x)$ are a straight line, as shown in figure 4.1, or a parabola. Both cases will be discussed next.

doi:10.1088/978-0-7503-6493-5ch4

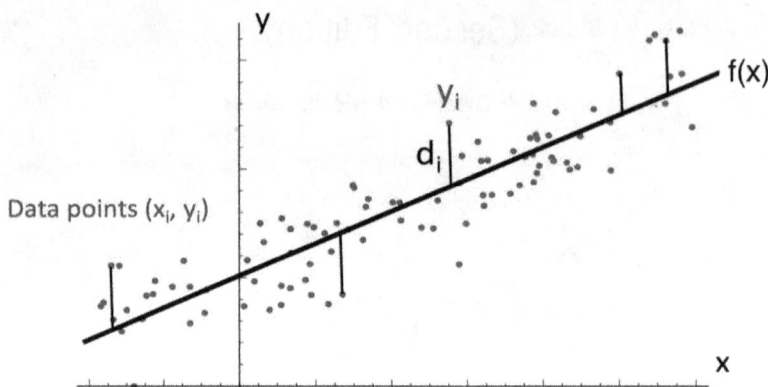

**Figure 4.1.** Experimental data points $(x_i, y_i)$ fitted by a straight line $f(x)$. The quantity $d_i$ defines the deviation of each data point $y_i$ from the best-fit straight line $f(x_i)$ at each data point $i$, that is, $d_i = y_i - f(x_i)$.

### 4.1.1 The linear least-squares approximation

The linear least-squares method uses a straight line, $y = a + bx$, to approximate a given set of data, $(x_1, y_1), (x_2, y_2), \cdots, (x_N, y_N)$, where $N \geqslant 2$. From equation (4.1) it then follows that the best-fit straight line $f(x)$ has the least square error

$$\Pi(a, b) \equiv \sum_{i=1}^{N} (y_i - f(x_i))^2 = \sum_{i=1}^{N} (y_i - (a + bx_i))^2, \tag{4.2}$$

which is to be minimized. Note that here $a$ and $b$ are unknown coefficients while all the data points $x_i$ and $y_i$ are given. The coefficients are found by minimizing $\Pi$ of equation (4.2), which leads to the conditions $\partial \Pi / \partial a = 0$ and $\partial \Pi / \partial b = 0$. From equation (4.2), these partial derivatives are given by

$$\frac{\partial \Pi(a, b)}{\partial a} = 2 \sum_{i=1}^{N} (y_i - (a + bx_i)) = 0, \tag{4.3}$$

$$\frac{\partial \Pi(a, b)}{\partial b} = 2 \sum_{i=1}^{N} x_i (y_i - (a + bx_i)) = 0. \tag{4.4}$$

Expanding equations (4.3) and (4.4), we obtain the following set of coupled equations (linear in $a$ and $b$),

$$a \sum_{i=1}^{N} 1 + b \sum_{i=1}^{N} x_i = \sum_{i=1}^{N} y_i, \tag{4.5}$$

$$a \sum_{i=1}^{N} x_i + b \sum_{i=1}^{N} x_i^2 = \sum_{i=1}^{N} x_i y_i. \tag{4.6}$$

Solving for $a$ and $b$ leads to

$$a = \frac{\left(\sum_{i=1}^{N} y_i\right)\left(\sum_{i=1}^{N} x_i^2\right) - \left(\sum_{i=1}^{N} x_i\right)\left(\sum_{i=1}^{N} x_i\, y_i\right)}{N\left(\sum_{i=1}^{N} x_i^2\right) - \left(\sum_{i=1}^{N} x_i\right)^2}, \tag{4.7}$$

$$b = \frac{\left(N\sum_{i=1}^{N} x_i\, y_i\right) - \left(\sum_{i=1}^{N} x_i\right)\left(\sum_{i=1}^{N} y_i\right)}{n\left(\sum_{i=1}^{N} x_i^2\right) - \left(\sum_{i=1}^{N} x_i\right)^2}, \tag{4.8}$$

which are easy to evaluate numerically.

### 4.1.2 The quadratic least-squares approximation

The quadratic least-squares method uses a second degree polynomial (i.e., a parabola) $y = a + bx + cx^2$ to approximate a given set of data, $(x_1, y_1)$, $(x_2, y_2)$, $\cdots$, $(x_N, y_N)$, where $N \geqslant 3$. From equation (4.1), it now follows that the best-fit polynomial $f(x)$ has the least square error

$$\Pi(a, b, c) \equiv \sum_{i=1}^{N} (y_i - f(x_i))^2 = \sum_{i=1}^{N} \left(y_i - (a + bx_i + cx_i^2)\right)^2 = \text{minimum.} \tag{4.9}$$

For a parabola, $a$, $b$, and $c$ are the unknown coefficients and all $x_i$ and $y_i$ data are given as for the straight-line case. To find the three unknown coefficients, we need to evaluate the three first-order derivatives

$$\frac{\partial \Pi(a, b, c)}{\partial a} = 2\sum_{i=1}^{N} \left(y_i - (a + bx_i + cx_i^2)\right) = 0, \tag{4.10}$$

$$\frac{\partial \Pi(a, b, c)}{\partial b} = 2\sum_{i=1}^{N} x_i \left(y_i - (a + bx_i + cx_i^2)\right) = 0, \tag{4.11}$$

$$\frac{\partial \Pi(a, b, c)}{\partial c} = 2\sum_{i=1}^{N} x_i^2 \left(y_i - (a + bx_i + cx_i^2)\right) = 0. \tag{4.12}$$

Expanding equations (4.10)–(4.12), we have

$$a\sum_{i=1}^{N} 1 + b\sum_{i=1}^{N} x_i + c\sum_{i=1}^{N} x_i^2 = \sum_{i=1}^{N} y_i, \tag{4.13}$$

$$a\sum_{i=1}^{N} x_i + b\sum_{i=1}^{N} x_i^2 + c\sum_{i=1}^{N} x_i^3 = \sum_{i=1}^{N} x_i\, y_i, \tag{4.14}$$

$$a\sum_{i=1}^{N} x_i^2 + b\sum_{i=1}^{N} x_i^3 + c\sum_{i=1}^{N} x_i^4 = \sum_{i=1}^{N} x_i^2 \, y_i. \tag{4.15}$$

The unknown coefficients $a$, $b$, and $c$ are given as solutions of the three coupled linear equations (4.13) through (4.15). To solve for the unknown coefficients $a$, $b$, and $c$, we express equations (4.13) through (4.15) in matrix form,

$$\begin{pmatrix} \sum_{i=1}^{N} 1 & \sum_{i=1}^{N} x_i & \sum_{i=1}^{N} x_i^2 \\ \sum_{i=1}^{N} x_i & \sum_{i=1}^{N} x_i^2 & \sum_{i=1}^{N} x_i^3 \\ \sum_{i=1}^{N} x_i^2 & \sum_{i=1}^{N} x_i^3 & \sum_{i=1}^{N} x_i^4 \end{pmatrix} \begin{pmatrix} a \\ b \\ c \end{pmatrix} = \begin{pmatrix} \sum_{i=1}^{N} y_i \\ \sum_{i=1}^{N} x_i y_i \\ \sum_{i=1}^{N} x_i^2 y_i \end{pmatrix}, \tag{4.16}$$

which allows us to use matrix algebra techniques to find the solution efficiently. Defining

$$\mathbf{A} \equiv \begin{pmatrix} N & \sum_{i=1}^{N} x_i & \sum_{i=1}^{N} x_i^2 \\ \sum_{i=1}^{N} x_i & \sum_{i=1}^{N} x_i^2 & \sum_{i=1}^{N} x_i^3 \\ \sum_{i=1}^{N} x_i^2 & \sum_{i=1}^{N} x_i^3 & \sum_{i=1}^{N} x_i^4 \end{pmatrix}, \quad \mathbf{x} \equiv \begin{pmatrix} a \\ b \\ c \end{pmatrix}, \quad \mathbf{b} \equiv \begin{pmatrix} \sum_{i=1}^{N} y_i \\ \sum_{i=1}^{N} x_i y_i \\ \sum_{i=1}^{N} x_i^2 y_i \end{pmatrix}, \tag{4.17}$$

the system of equations can then be written in the form

$$\mathbf{A}\mathbf{x} = \mathbf{b}, \tag{4.18}$$

where $\mathbf{A}$ is the matrix of coefficients, $\mathbf{x}$ is the vector of unknowns, and $\mathbf{b}$ is the vector of constants on the right-hand side of equations (4.13)–(4.15). To solve the system $\mathbf{A}\mathbf{x} = \mathbf{b}$, one can utilize the matrix inversion method [1, 2], provided that the matrix $\mathbf{A}$ is invertible. The solution is expressed as $\mathbf{x} = \mathbf{A}^{-1}\mathbf{b}$. Numerical libraries such as LAPACK, LINPACK, IMSL, or NAG [3–6] provide very efficient numerical codes to compute inverse matrices.

Finally, to outline how the equations for a higher-order polynomial fit look, let us consider a fourth-order polynomial of the form

$$y = a_0 + a_1 x + a_2 x^2 + a_3 x^3 + a_4 x^4. \tag{4.19}$$

In this case, $\mathbf{A}$ is a $5 \times 5$ matrix given by

$$\mathbf{A} = \begin{pmatrix} N & \sum_{i=1}^{N} x_i & \sum_{i=1}^{N} x_i^2 & \sum_{i=1}^{N} x_i^3 & \sum_{i=1}^{N} x_i^4 \\ \sum_{i=1}^{N} x_i & \sum_{i=1}^{N} x_i^2 & \sum_{i=1}^{N} x_i^3 & \sum_{i=1}^{N} x_i^4 & \sum_{i=1}^{N} x_i^5 \\ \sum_{i=1}^{N} x_i^2 & \sum_{i=1}^{N} x_i^3 & \sum_{i=1}^{N} x_i^4 & \sum_{i=1}^{N} x_i^5 & \sum_{i=1}^{N} x_i^6 \\ \sum_{i=1}^{N} x_i^3 & \sum_{i=1}^{N} x_i^4 & \sum_{i=1}^{N} x_i^5 & \sum_{i=1}^{N} x_i^6 & \sum_{i=1}^{N} x_i^7 \\ \sum_{i=1}^{N} x_i^4 & \sum_{i=1}^{N} x_i^5 & \sum_{i=1}^{N} x_i^6 & \sum_{i=1}^{N} x_i^7 & \sum_{i=1}^{N} x_i^8 \end{pmatrix}, \tag{4.20}$$

and the fifth-dimensional vectors $\mathbf{x}$ and $\mathbf{b}$ are given by

$$\mathbf{x} = \begin{pmatrix} a_0 \\ a_1 \\ a_2 \\ a_3 \\ a_4 \end{pmatrix}, \quad \mathbf{b} = \begin{pmatrix} \sum_{i=1}^{N} y_i \\ \sum_{i=1}^{N} x_i y_i \\ \sum_{i=1}^{N} x_i^2 y_i \\ \sum_{i=1}^{N} x_i^3 y_i \\ \sum_{i=1}^{N} x_i^4 y_i \end{pmatrix}. \tag{4.21}$$

The generalization to a polynomial of order $M$

$$y = a_0 + a_1 x + a_2 x^2 + a_3 x^3 + \ldots + a_{M-1} x^{M-1} + a_M x^M \tag{4.22}$$

results in the matrix $\mathbf{A}$ being of dimension $(M+1) \times (M+1)$, and the vectors $\mathbf{x}$ and $\mathbf{b}$ each being of dimension $M+1$.

## 4.2 The cubic spline approximation

Splines are piecewise-defined polynomials that provide a smooth and accurate fit to data [7–9]. The most commonly used splines are cubic splines, which are piecewise third-degree polynomials. Cubic splines are constructed such that the interpolating function is continuous and has continuous first and second derivatives, ensuring a smooth transition between polynomial segments. The mathematical framework underlying (natural) cubic splines will be presented in this section. A complete Fortran code which solves these equations for the experimental data of table 4.1 is

**Table 4.1.** Experimental data for $x$ and $y$.

| i | 1 | 2 | 3 | 4 | 5 | 6 | 7 | 8 | 9 | 10 |
|---|---|---|---|---|---|---|---|---|---|---|
| x(i) | 0.0 | 0.5 | 1.0 | 1.5 | 2.0 | 2.5 | 3.0 | 3.5 | 4.0 | 4.5 |
| y(i) | 0.0 | 1.0 | 0.5 | 2.0 | 1.5 | 0.7 | 1.2 | 2.5 | 1.8 | 1.0 |

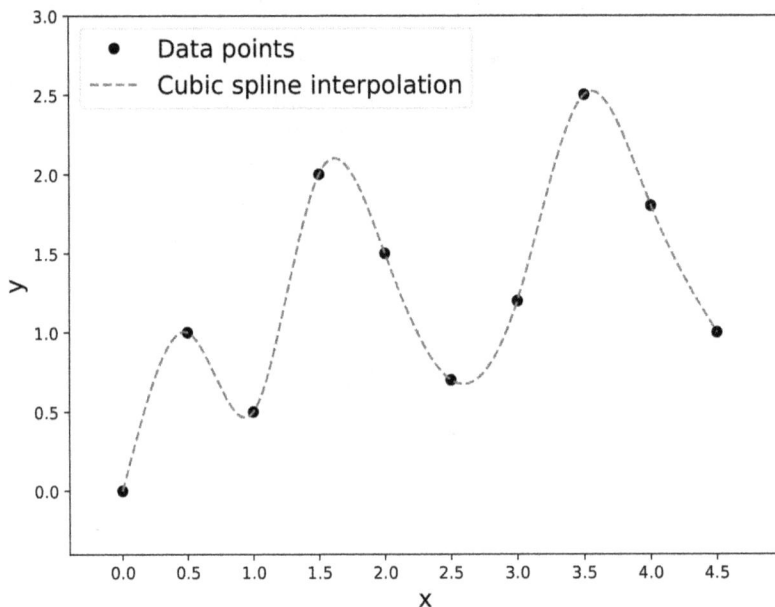

**Figure 4.2.** Graphical illustration of the data shown in table 4.1 fitted with a cubic spline.

shown in appendix A. Figure 4.2 shows the experimental data along with the cubic spline fit curve produced by the Fortran code shown in appendix A.

The number of data points in table 4.1 is just 10. Let us assume we have not 10 but a set of $n$ data points $(x_0, y_0)$, $(x_1, y_1)$, ... , $(x_n, y_n)$. The goal is to find a cubic spline $S(x)$ such that $S(x_i) = y_i$ for $i = 0, 1, ... , n$. Each segment of the cubic spline $S_i(x)$ between $x_i$ and $x_{i+1}$ is a cubic polynomial,

$$S_i(x) = a_i + b_i(x - x_i) + c_i(x - x_i)^2 + d_i(x - x_i)^3. \tag{4.23}$$

The coefficients $a_i$, $b_i$, $c_i$, and $d_i$ are determined by solving a system of equations that enforce the continuity and smoothness conditions at the data points. They are determined in four steps, which are as follows:

Step 1: Compute intervals and coefficients,

$$h_i = x_{i+1} - x_i, \quad a_i = y_i \tag{4.24}$$

Step 2: Compute the coefficients $\alpha_i$,

$$\alpha_i = \frac{3}{h_i}(a_{i+1} - a_i) - \frac{3}{h_{i-1}}(a_i - a_{i-1}), \quad i = 2, \ldots, n - 1, \qquad (4.25)$$

Step 3: Solve the tridiagonal system,

$$l_1 = 1, \quad \mu_1 = 0, \quad z_1 = 0, \qquad (4.26)$$

$$l_i = 2(x_{i+1} - x_{i-1}) - h_{i-1}\mu_{i-1}, \quad i = 2, \ldots, n - 1, \qquad (4.27)$$

$$\mu_i = \frac{h_i}{l_i}, \quad z_i = \frac{\alpha_i - h_{i-1}z_{i-1}}{l_i}, \quad i = 2, \ldots, n - 1, \qquad (4.28)$$

$$l_n = 1, \quad z_n = 0, \quad c_n = 0. \qquad (4.29)$$

Backward substitution to solve for $c_j$,

$$c_j = z_j - \mu_j c_{j+1}, \quad j = n - 1, \ldots, 1. \qquad (4.30)$$

Step 4: Compute the spline coefficients $b_j$ and $d_j$,

$$b_j = \frac{a_{j+1} - a_j}{h_j} - \frac{h_j}{3}(c_{j+1} + 2c_j), \quad j = 1, \ldots, n - 1, \qquad (4.31)$$

$$d_j = \frac{c_{j+1} - c_j}{3h_j}, \quad j = 1, \ldots, n - 1. \qquad (4.32)$$

For each $x$ in the interpolation range, the spline is evaluated as

$$S_i(x) = a_j + b_j(x - x_j) + c_j(x - x_j)^2 + d_j(x - x_j)^3 \quad \text{for} \quad x_j \leqslant x \leqslant x_{j+1} \qquad (4.33)$$

## 4.3 Numerical differentiation

Derivatives of smooth, well-behaved functions can be approximated in several ways. The most basic ones are discussed in this section. We begin with the Taylor series expansion of a function $f(x)$,

$$f(x + \Delta x) = f(x) + f'(x)\Delta x + \frac{f''(x)}{2!}\Delta x^2 + \frac{f'''(x)}{3!}\Delta x^3 + \cdots, \qquad (4.34)$$

where $\Delta x \ll x$. Evaluating the Taylor expansion (4.34) at $x + \Delta x$ and $x - \Delta x$ leads to

$$f(x + \Delta x) \approx f(x) + f'(x)\Delta x + \frac{f''(x)}{2!}\Delta x^2 + \frac{f'''(x)}{3!}\Delta x^3, \qquad (4.35)$$

$$f(x - \Delta x) \approx f(x) - f'(x)\Delta x + \frac{f''(x)}{2!}\Delta x^2 - \frac{f'''(x)}{3!}\Delta x^3. \qquad (4.36)$$

Subtracting equation (4.36) from (4.35) gives

$$f(x + \Delta x) - f(x - \Delta x) \approx 2f'(x)\Delta x + 2\frac{f'''(x)}{3!}\Delta x^3, \tag{4.37}$$

so that

$$f'(x) \approx \frac{f(x + \Delta x) - f(x - \Delta x)}{2\Delta x}, \tag{4.38}$$

which is known as leap-frog (central difference) differentiation. A less accurate way of numerical differentiation is the so-called Euler forward differentiation, which follows from equation (4.35) as

$$f'(x) \approx \frac{f(x + \Delta x) - f(x)}{\Delta x}. \tag{4.39}$$

Finally, we mention Euler backward differentiation, which follows from equation (4.36) as

$$f'(x) \approx \frac{f(x) - f(x - \Delta x)}{\Delta x}. \tag{4.40}$$

A frequently used expression to compute the second-order derivative of a function $f(x)$ is obtained by adding equations (4.35) and (4.36) together. This leads to the three-point rule formula

$$f''(x) \approx \frac{f(x + \Delta x) - 2f(x) + f(x - \Delta x)}{\Delta x^2}. \tag{4.41}$$

Similarly, the five-point rule formula for the second derivative of $f(x)$ is given by

$$f''(x) \approx \frac{-f(x + 2h) + 16f(x + h) - 30f(x) + 16f(x - h) - f(x - 2h)}{12h^2}. \tag{4.42}$$

## 4.4 Numerical integration

There are several excellent methods that can be used to numerically compute the value $w$ of a definite integral of the form

$$w = \int_a^b f(x)dx. \tag{4.43}$$

Here, $f(x)$ is a smooth, well-behaved function defined for $x$-values in the range $a \leqslant x \leqslant b$, as illustrated in figure 4.3. A large class of numerical integration (quadrature) schemes can be derived by constructing interpolating functions which are easy to integrate.

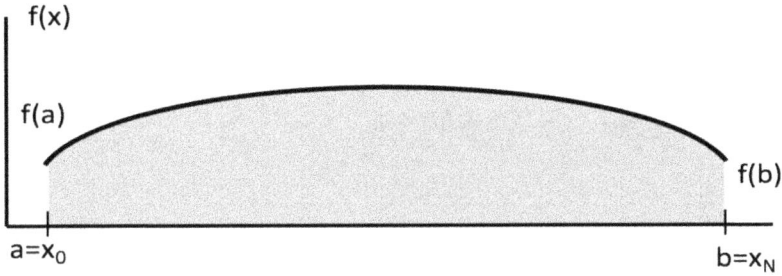

**Figure 4.3.** Graphical illustration of the value $w$ (shaded area) of the integral of equation (4.43).

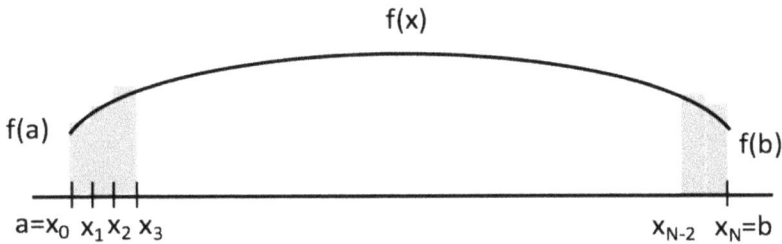

**Figure 4.4.** The the trapezoidal rule, the area covered by $f(x)$ is approximated by a sequence of vertical trapezoids of width $x_{i+1} - x_i$.

## 4.4.1 The trapezoidal rule

In the simplest case, the functions used to interpolate $f(x)$ are polynomials of degree one, i.e., linear functions, which are used to approximate the area under a function $f(x)$ for $x$-values between $x_i$ and $x_{i+1}$ by $N$ trapezoids, as shown in figure 4.4. Here, $N$ is an integer which can be even or odd. Defining

$$x_k = a + kh, \quad \text{where} \quad h = (b - a)/N, \quad (k = 0, 1, \ldots, N) \quad (4.44)$$

shows that the difference between two successive grid points on the $x$-axis is given by $x_{i+1} - x_i = h$. Since the area of each individual trapezoid is $hf(x_k) + h(f(x_{k+1}) - f(x_k))/2$, the total area covered by $f(x)$ follows as

$$\int_a^b f(x)dx \approx \sum_{k=0}^{N-1} hf(x_k) + \frac{1}{2}\sum_{k=0}^{N-1} h(f(x_{k+1}) - f(x_k)). \quad (4.45)$$

This expression can be written in the more compact form

$$\int_a^b f(x)dx \approx \sum_{k=0}^{N-1} hf_k + \frac{1}{2}\sum_{k=0}^{N-1} h(f_{k+1} - f_k), \quad (4.46)$$

where $f_k \equiv f(x_k)$ and $f_{k+1} \equiv f(x_{k+1})$. Next, we rewrite the summations in equation (4.46) as follows,

$$\sum_{k=0}^{N-1} hf_k + \frac{1}{2}\sum_{k=0}^{N-1} h(f_{k+1} - f_k) = hf_0 + h\sum_{k=1}^{N-1} f_k + \frac{h}{2}\sum_{k=0}^{N-1} f_{k+1} - \frac{h}{2}\sum_{k=0}^{N-1} f_k, \quad (4.47)$$

and furthermore

$$\sum_{k=0}^{N-1} f_{k+1} = \sum_{k=1}^{N-1} f_k + f_N, \quad (4.48)$$

$$\sum_{k=0}^{N-1} f_k = f_0 + \sum_{k=1}^{N-1} f_k. \quad (4.49)$$

This allows us to write the expression for the integral in equation (4.46) as

$$\int_a^b f(x)dx \approx \frac{h}{2}\left( f_a + 2\sum_{k=1}^{N-1} f_k + f_b \right), \quad (4.50)$$

with $f_a \equiv f(a)$, $f_b \equiv f(b)$, and $x_k$ and $h$ defined in equation (4.44). Equation (4.50) is known as the trapezoidal rule. The sample code below illustrates how the trapezoidal rule can be implemented in a numerical code:

```
h=(b-a)/FLOAT(N)
sum=0.
DO k=1,N-1
        x_k=a+h*FLOAT(k)
        sum=sum+f(x_k)
END DO
trap = h * ( f(a)+f(b)+2.*sum ) / 2.
```

### 4.4.2 Simpson's rule

The second numerical integration technique considered in this text is Simpson's rule, which is often more accurate than the trapezoidal rule because it uses a second-order polynomial (i.e., a parabola) $P(x) = Ax^2 + Bx + C$ rather than a linear function to approximate $f(x)$ between grid points $x_i$ and $x_{i+2}$. The situation is illustrated graphically in figure 4.5. The quantities $A$, $B$, and $C$ are constants which can be determined by integrating $P(x)$ from $x_1$ to $x_{i+2}$,

$$\int_{x_i}^{x_{i+2}} P(x)dx = \frac{x_{i+2} - x_i}{3}\left( A\left(x_{i+2}^2 + x_{i+2}x_i + x_i^2\right) + \frac{3}{2}B(x_{i+2} + x_i) + 3C \right). \quad (4.51)$$

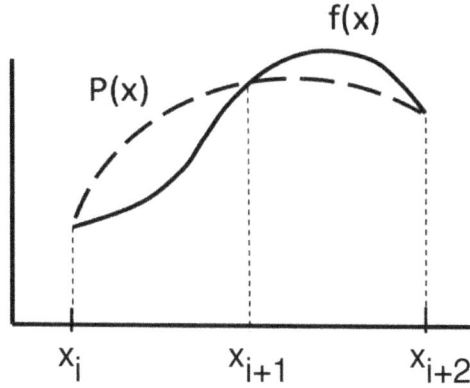

**Figure 4.5.** Simpson's rule can be derived by approximating the integrand $f(x)$ by a quadratic interpolant $P(x)$ between grid points $x_i$ and $x_{i+2}$.

With a little bit of algebra, equation (4.51) can be written as

$$\int_{x_i}^{x_{i+2}} P(x)dx = \frac{x_{i+2} - x_i}{6}\left(P(x_i) + 4P\left(\frac{x_i + x_{i+2}}{2}\right) + P(x_{i+2})\right). \qquad (4.52)$$

Now let us suppose that the integration interval $[a, b]$ is split up into $N$ sub-intervals, where $N$ is an even integer. Then, by applying the rule derived in equation (4.52) to integrate $f(x)$ over the first two intervals $a, x_2$ (see figure 4.4) leads to $w_2 \equiv (f_0 + 4f_1 + f_2)/3$, where $h$ denotes the width of the interval, i.e., $h = x_2 - a$. Repeating this step for all subsequent pairs of adjacent intervals leads for the integral to

$$\int_a^b f(x)dx \approx \frac{h}{3}(f_0 + 4f_1 + f_2) + \frac{h}{3}(f_2 + 4f_3 + f_4) + \frac{h}{3}(f_4 + 4f_5 + f_6)$$

$$+ \ldots + \frac{h}{3}(f_{N-2} + 4f_{N-1} + f_N), \qquad (4.53)$$

which can be written as

$$\int_a^b f(x)dx \approx \frac{h}{3}(f_0 + f_N) + \frac{h}{3}(f_0 + 2f_2 + 2f_4 + \ldots + 2f_{N-2} + f_N)$$

$$+ \frac{h}{3}(4f_1 + 4f_3 + 4f_6 + \ldots + 4f_{N-1}). \qquad (4.54)$$

One sees that the $f_k$ values get multiplied by a factor of 2 or 4 depending on whether the grid point is even or odd, respectively. This is different for the trapezoidal rule of equation (4.50), where all $f_k$ values are multiplied by the same weight factor. In summary, Simpson's rule is given by

$$\int_a^b f(x)dx \approx \frac{h}{3}\left( f_a + 2\sum_{\substack{k=2 \\ \uparrow \\ \text{even}}}^{N-2} f_k + 4\sum_{\substack{k=1 \\ \uparrow \\ \text{odd}}}^{N-1} f_k + f_b \right), \tag{4.55}$$

where $f_a \equiv f_0$ and $f_b \equiv f_N$. This formula can also be written as

$$\int_a^b f(x)dx \approx \frac{h}{3}\left( f_0 + \sum_{k=1}^{N-1}(2\delta_{(-1)^k,+1} + 4\delta_{(-1)^k,-1})f_k + f_N \right), \tag{4.56}$$

where the Kronecker delta has been used to specify the corresponding weight factor, which is either 2 for even $k$ values or 4 for odd $k$ values. The Kronecker $\delta_{a,b}$ has a value of 1 if $a = b$ and 0 if $a \neq b$. As is the case for the trapezoidal rule (see equation (4.44)), the grid points are given by $x_k = a + kh$, where $k = 0, 1, \dots, N-1, N$, and $h = (b - a)/N$. The sample code below shows how Simpson's rule can be implemented in a Fortran code, where `sign=(-1)**k` is used to discriminate between even and odd grid points:

```
h=(b-a)/FLOAT(N)
sum=0.
DO k=1,N-1
        x_k=a+h*FLOAT(k)
        weight=4.
        sign=(-1)**k
        if(sign > 0) weight=2.
        sum=sum+weight*f(x_k)
END DO
simp = h * (f(a)+f(b)+sum) / 3.
```

Fortran offers the mod function, which can also be used to alternate between the weight factors 2 and 4 for even and odd values of $k$, respectively. The usage is mod (k, 2), which returns 0 if $k$ is even and 1 if $k$ is odd.

## 4.5 Monte Carlo integration

Monte Carlo (MC) methods are a broad class of computational algorithms that rely on repeated random sampling to obtain numerical results. The name 'Monte Carlo' is a reference to the Monte Carlo Casino in Monaco, reflecting the element of chance at the heart of these methods. They are widely used in various fields such as physics, finance, and engineering to model complex systems and solve problems that are difficult or impossible to solve analytically. MC techniques are particularly powerful for solving integrals, optimizing functions, and simulating physical systems. They leverage the law of large numbers, which states that the average of the results obtained from a large number of trials should be close to the expected value.

### 4.5.1 Mathematical background of MC integration

MC integration is a method for numerically estimating the value of an integral using random sampling. This technique is particularly useful for high-dimensional integrals and complex domains where traditional numerical integration methods become impractical. Consider the problem of estimating the definite integral of a function $f(x)$ over a domain $D$,

$$I = \int_D f(x)\, dx. \tag{4.57}$$

For simplicity, let us consider a one-dimensional integral over an interval $[a, b]$,

$$I = \int_a^b f(x)\, dx. \tag{4.58}$$

MC integration approximates this integral by randomly sampling points within the domain and averaging the function values at these points. A computer code thus first needs to generate $N$ random points $x_i$ uniformly distributed over the interval $[a, b]$. The function $f(x)$ is then evaluated at each sampled point $x_i$, followed by computing the average of these function values. Finally, one needs to multiply this average by the length of the interval $(b - a)$. Mathematically, this can be expressed as:

$$I \approx \frac{b-a}{N} \sum_{i=1}^{N} f(x_i). \tag{4.59}$$

A practical example is illustrated next, showing how to integrate $f(x) = x^2$ over the interval $[0, 1]$, i.e.,

$$I = \int_0^1 x^2\, dx = \frac{1}{3}. \tag{4.60}$$

As outlined above, the following steps need to be taken successively:
- Generate $N$ random points $x_i$ uniformly distributed in $[0, 1]$.
- Evaluate $f(x) = x_i^2$ for each $x_i$.
- Compute the estimate by averaging and scaling, i.e.,

$$I \approx \frac{1}{N} \sum_{i=1}^{N} x_i^2$$

method can converge even if the matrix is not strictly diagonally dominant. To illustrate, suppose we have the following linear system:

$$
\begin{aligned}
4x_1 + x_2 + x_3 &= 7, \\
x_1 + 5x_2 + x_3 &= -8, \\
x_1 + x_2 + 6x_3 &= 6.
\end{aligned}
\tag{4.76}
$$

The corresponding matrix $\mathbf{A}$ is:

$$
\mathbf{A} = \begin{pmatrix} 4 & 1 & 1 \\ 1 & 5 & 1 \\ 1 & 1 & 6 \end{pmatrix}.
\tag{4.77}
$$

To check for diagonal dominance:

$$
\begin{aligned}
|4| &\geqslant |1| + |1|, \\
|5| &\geqslant |1| + |1|, \\
|6| &\geqslant |1| + |1|.
\end{aligned}
\tag{4.78}
$$

Since each diagonal element is greater than or equal to the sum of the off-diagonal elements in its row, the matrix is diagonally dominant. Therefore, the Jacobi method will converge for this system. For matrices that do not meet these criteria, other iterative methods, such as the Gauss–Seidel method [10] or successive over-relaxation [11], might be more appropriate because they can have better convergence properties.

## 4.7 Finding roots

The objective of this section is to introduce a numerical tool for finding theroot of a transcendental equation $f(x)$, i.e., the $\tilde{x}$ value for which $f(\tilde{x}) = 0$. A brute force method is the so-called bisection method, which cuts the interval in half in which the root lies, as shown in figure 4.6. This procedure is repeated until $f(x)$, evaluated at every new midpoint, is smaller than a prescribed tolerance. A more elegant and

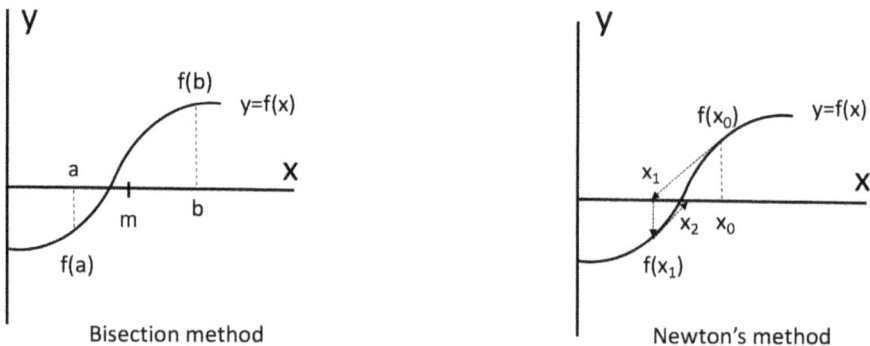

**Figure 4.6.** Left-hand panel: The bisection method determines the midpoint of the interval in which the root of $f(x)$ lies. Right-hand panel: Newton's method uses line tangents to $f$ to find the root of $f(x)$.

efficient method to find the root of a function is Newton's method. Here, one starts from a given initial guess value, $x_0$, for the root. A refined value for the guess value, $x_1$, is computed based on the $x$-intercept of the line tangent to $f$ at $x_0$, as shown in figure 4.6. Mathematically, this is expressed as

$$(f(x_1) - f(x_0)) = f'(x_0)\,(x_1 - x_0). \tag{4.79}$$

Since $f(x_1) = 0$ by construction, it follows from equation (4.79) that

$$0 - f(x_0) = f'(x_0)\,(x_1 - x_0) \quad \Rightarrow \quad x_1 = x_0 - \frac{f(x_0)}{f'(x_0)}, \tag{4.80}$$

where $x_1$ is the new, improved value for root. This process is repeated according to the scheme ($i \in \mathbb{N}_+$)

$$x_{i+1} = x_i - \frac{f(x_i)}{f'(x_i)}, \tag{4.81}$$

until an $x_{i+1}$ value is found for which $|f(x_{i+1})| < \varepsilon$, where $\varepsilon$ is a prescribed tolerance.

## 4.8 Solving ordinary differential equations

An ordinary differential equation (ODE) of order $n$ is an equation of the form

$$F(x; y, y^{(1)}, y^{(2)}, \ldots, y^{(n)}) = 0, \tag{4.82}$$

where $y$ is a function of $x$, $y^{(1)} = dy/dx$, and $y^{(n)} = d^n y/dx^n$. In general, equation (4.82) has $n$ linearly independent solutions, which are determined by initial conditions $Y_0 \equiv y(0)$, $Y_1 \equiv y^{(1)}(0)$, $\cdots$, $Y_{n-1} \equiv y^{(n-1)}(0)$. In general, higher-order ODEs can be reduced to a set of first-order ODEs, which are easier to solve numerically. To demonstrate how this works, let us consider the second-order OED

$$P(x)\,\frac{d^2 y(x)}{dx^2} + Q(x)\,\frac{dy(x)}{dx} = S(x), \tag{4.83}$$

where $P(x)$, $Q(x)$, and $S(x)$ are given function of $x$. This equation has two linearly independent solutions, $y_1(x)$ and $y_2(x)$, with initial conditions $Y_0$ and $Y_1$. The functions $y_1(x)$ and $y_2(x)$ can be computed from the following set of coupled, first-order differential equations

$$\frac{dy_1(x)}{dx} = y_2(x), \tag{4.84}$$

$$\frac{dy_2(x)}{dx} = \frac{1}{P(x)}\,(S(x) - Q(x)\,y_2(x)), \tag{4.85}$$

which is equivalent to equation (4.83), as can be seen by setting $y_1(x) = y(x)$ and $y_2(x) = dy_1/dx$. Similarly, a linear $n^{\text{th}}$-order ODE can be reduced to a system of $n$ coupled first-order ODEs, according to the scheme

$$\frac{dy_k(x)}{dx} = f_k(y_1(x), y_2(x), \ldots, y_n(x)), \quad (k = 1, 2, \ldots, n), \tag{4.86}$$

with initial conditions $Y_0$, $Y_1$, $\cdots$, $Y_{n-1}$. Common numerical methods for solving initial value problems of ODEs are the Euler method, the midpoint method, and the Runge–Kutta method. These methods will be introduced next.

### 4.8.1 The Euler method

The Euler method is the simplest method to approximate first-order derivatives. It advances the solution of a first-order ODE of the form $y' = f(x, y(x))$ (where $y' \equiv dy/dx$) with a given initial condition $Y_0 = y(0)$ from $x$ to $x + h$ according to

$$y(x + h) = y(x) + h f(x, y(x)). \tag{4.87}$$

Here, $h$ is a properly chosen, small marching step which advances the solution form $x$ to $x + h$. To indicate that the values of $x$ are discretized numerically, it is convenient to write equation (4.87) as

$$y(x_{n+1}) = y(x_n) + h f(x_n, y(x_n)) \tag{4.88}$$

$$= y(x_n) + k_1, \tag{4.89}$$

where $k_1 \equiv h f(x_n, y(x_n))$. Euler's formula (4.89) can easily be solved numerically, but it is limited in practical usage since it uses the derivative information only at the beginning of an interval, which implies numerical errors that may grow quickly, depending on the behavior of $y(x)$.

To illustrate how Euler's method is used to solve a second-order differential equation, let us consider a small object of mass $m$ which moves vertically ($z$-direction) through a viscous medium. The medium exerts a frictional force on the mass described by $b\,\dot{z}|\dot{z}|$, where $b$ is a constant and the speed of the mass is given by $\dot{z} = dz/dt$. The gravitational force acting on the sphere is given by $-m\,g$, with $g$ denoting the gravitational acceleration. The equation of motion of the mass follows from Newton's law and is given by

$$m\ddot{z} = -mg - b\dot{z}|\dot{z}|. \tag{4.90}$$

Let us assume that at the initial time, $t = 0$, the mass is at $z(0) = 0$ and its initial velocity is $\dot{z}(0) = 10$ m s$^{-1}$, which corresponds to $Y_0 = 0$ and $Y_1 = 10$ m s$^{-1}$, respectively. The system of first-order ODEs associated with equation (4.90) is obtained as follows,

$$v = \dot{z} \Rightarrow z(t + \Delta t) = z(t) + v(t)\,\Delta t, \tag{4.91}$$

$$\dot{v} = \ddot{z} \Rightarrow v(t + \Delta t) = v(t) - \frac{1}{m}(m\,g + b\,v(t)\,|v(t)|)\Delta t, \tag{4.92}$$

where $\Delta t$ is a properly chosen, small time step which advances the solution $z(t)$ form $t$ to $t + \Delta t$. Since the equations are solved for discretized times and positions, it is appropriate to write equations (4.91) and (4.92) as

$$z_{i+1} = z_i + v_i \, \Delta t, \tag{4.93}$$

$$v_{i+1} = v_i - \frac{1}{m}(m\,g + b\,v_i\,|v_i|)\,\Delta t. \tag{4.94}$$

A numerical sample code which solves equations (4.93) and (4.94) subject to the boundary conditions $Y_0 = 0$ and $Y_1 = 10$ m s$^{-1}$ is shown below.

```
z_1 = 0.0; v_1 = 10.0 !Define initial conditions
DO WHILE (z_2 >= 0.0)
      z_2 = z_1 + v_1 * dt
      v_2 = v_1 - dt * ( m*g + b*v_1*ABS(v_1) ) / m
      time = time + dt
END DO
```

### 4.8.2 The midpoint method

The midpoint method, also known as the second-order Runge–Kutta method, improves the Euler method by adding a midpoint in the step, which increases the numerical accuracy. Equation (4.89) is then replaced by

$$y(x_{n+1}) = y(x_n) + k_2, \tag{4.95}$$

where $k_1$ and $k_2$ are given by

$$k_1 = h f(x_n, y_n), \tag{4.96}$$

$$k_2 = h f\left(x_n + \frac{h}{2}, y(x_n) + \frac{k_1}{2}\right). \tag{4.97}$$

### 4.8.3 The Runge–Kutta method

The Runge–Kutta method solves an ODE of the form $y' = f(x, y(x))$ by determining the value of $y(x + h)$ in terms of $y(x)$ computed at several different $x$-values. To illustrate the method, let us begin with writing $y(x + h)$ as

$$y(x + h) = y(x) + (y(x + h) - y(x)) \tag{4.98}$$

$$= y(x) + \int_x^{x+h} y'(s)\, ds, \tag{4.99}$$

where $y' = dy/ds$. Defining $s = x + \tau h$ so that $ds = h\, d\tau$. This leads for the integral in equation (4.99) to

$$\int_x^{x+h} y'(s)\, ds = \int_0^1 y'(x + \tau h)\, d\tau, \tag{4.100}$$

so that equation (4.99) can be written as

$$y(x + h) = y(x) + h \int_0^1 y'(x + \tau h)\, d\tau. \tag{4.101}$$

Next we approximate the integral in equation (4.101) by a finite sum,

$$\int_0^1 y'(x + \tau h)\, d\tau = \sum_{i=1}^m b_i\, y'(x + c_i h), \tag{4.102}$$

where the $b_i$ denote unknown expansion coefficients. For y' $\equiv 1$ the values of these coefficients are constraint by the condition

$$\sum_{i=1}^m b_i = 1. \tag{4.103}$$

Substituting equation (4.102) into equation (4.101) leads to

$$y(x + h) = y(x) + h \sum_{i=1}^m b_i\, y'(x + c_i h) \tag{4.104}$$

$$= y(x) + h \sum_{i=1}^m b_i\, f(x + c_i h,\, y(x + c_i h)). \tag{4.105}$$

To evaluate equation (4.105) further, we need to find an approximate expression for $y$ at the new grip points $x + c_i h$. With this in mind, we make use of equation (4.101) to arrive for $y(x + c_i h)$ at

$$y(x + c_i h) = y(x) + h \int_0^{c_i} y'(x + \tau h)\, d\tau. \tag{4.106}$$

Approximating the integral in equation (4.106) by a finite sum, as in equation (4.102), leads to ($i = 1,\, \ldots,\, m,\, j = 1,\, \ldots,\, m$)

$$\int_0^{c_i} y'(x + \tau h)\, d\tau = \sum_{j=1}^m a_{i,j}\, y'(x + c_j h). \tag{4.107}$$

As before, for $y' \equiv 1$ we obtain the conditions

$$\sum_{j=1}^m a_{i,j} = c_i. \tag{4.108}$$

Substituting equation (4.107) into equation (4.106) leads to

$$y(x + c_i h) = y(x) + h\sum_{j=1}^{m} a_{i,j}\, y'(x + c_j h) \tag{4.109}$$

$$= y(x) + h\sum_{j=1}^{m} a_{i,j}\, f(x + c_j h,\, y(x + c_j h)). \tag{4.110}$$

Finally, in order to simplify the notation, we introduce the abbreviation

$$\tilde{k}_j \equiv f(x + c_j h,\, y(x + c_j h)). \tag{4.111}$$

equation (4.110) can then be written in the more compact form

$$y(x + c_i h) = y(x) + h\sum_{j=1}^{m} a_{i,j}\, \tilde{k}_j, \quad (i = 1, \dots, m). \tag{4.112}$$

Plugging equation (4.112) back into equation (4.111) leads to

$$\tilde{k}_j = f\left(x + c_j h,\, y(x) + h\sum_{l=1}^{m} a_{j,l}\, \tilde{k}_l\right), \quad (j = 1, \dots, m). \tag{4.113}$$

The set of Runge–Kutta equations of order $m$ follows from equation (4.105), which, by means of equations (4.111) and (4.113), can be written in the final form

$$y_{n+1} = y_n + h\sum_{i=1}^{m} b_i\, \tilde{k}_i, \tag{4.114}$$

$$\tilde{k}_i = f\left(x_n + c_i h,\, y_n + h\sum_{j=1}^{m} a_{i,j}\, \tilde{k}_j\right), \tag{4.115}$$

where $y_n \equiv y(x_n)$ and $y_{n+1} \equiv y(x_n + h)$. For $m = 4$, the unknown coefficients $c_i$, $b_i$, and $a_{i,j}$, summarized schematically as

$$\begin{array}{ccccc}
c_1 & a_{1,1} & a_{1,2} & a_{1,3} & a_{1,4} \\
c_2 & a_{2,1} & a_{2,2} & a_{2,3} & a_{2,4} \\
c_3 & a_{3,1} & a_{3,2} & a_{3,3} & a_{3,4} \\
c_4 & a_{4,1} & a_{4,2} & a_{4,3} & a_{4,4} \\
 & b_1 & b_2 & b_3 & b_4
\end{array} \tag{4.116}$$

have the following values,

$$\begin{array}{ccccc}
0 & 0 & 0 & 0 & 0 \\
1/2 & 1/2 & 0 & 0 & 0 \\
1/2 & 0 & 1/2 & 0 & 0 \\
1 & 0 & 0 & 1 & 0 \\
 & 1/6 & 2/6 & 2/6 & 1/6
\end{array} \tag{4.117}$$

This leads for $\tilde{k}_1$, $\tilde{k}_2$, $\tilde{k}_3$, and $\tilde{k}_4$ to

$$\tilde{k}_1 = f(x_n, y_n),$$
$$\tilde{k}_2 = f(x_n + \frac{1}{2}h, y_n + \frac{1}{2}h\tilde{k}_1),$$
$$\tilde{k}_3 = f(x_n + \frac{1}{2}h, y_n + \frac{1}{2}h\tilde{k}_2), \qquad (4.118)$$
$$\tilde{k}_4 = f(x_n + h, y_n + h\tilde{k}_3),$$
$$y_{n+1} = y_n + \frac{1}{6}h(\tilde{k}_1 + 2\tilde{k}_2 + 2\tilde{k}_3 + \tilde{k}_4).$$

It is convenient to define $k_i \equiv h\tilde{k}_i$, the equations of the fourth-order Runge–Kutta method, which is by far the most common method to solve ODE, can be summarized as follows:

$$k_1 = hf(x_n, y_n),$$
$$k_2 = hf(x_n + \frac{h}{2}, y(x_n) + \frac{k_1}{2}),$$
$$k_3 = hf(x_n + \frac{h}{2}, y(x_n) + \frac{k_2}{2}), \qquad (4.119)$$
$$k_4 = hf(x_n + h, y(x_n) + k_3),$$
$$y_{n+1} = y_n + \frac{1}{6}(k_1 + 2k_2 + 2k_3 + k_4).$$

### 4.8.4 Boundary value problems

In the previous sections, we looked at ODEs whose solutions and derivatives have specific values at given points, such as position and velocity at an initial time. Such problems are referred to as initial value problems. This is different for so-called boundary value problems, where the solutions are required to have specific values at the boundaries of the system that is being studied.

As an example, let us consider a particle of mass $m$ moving along the $x$-axis under the action of a time-dependent force $F(t)$. At time zero the particle is located at $x(0) = 0$. At the final time, $t_f$, the particle is at $x(t_f) = b$. The particle's motion is therefore described by the solution to the boundary value problem

$$m\frac{d^2x}{dt^2} = F(t), \quad \text{where} \quad x(0) = 0, \quad x(t_f) = b. \qquad (4.120)$$

To determine the motion of the particle from $x(0) = 0$ to $x(t_f) = a$ numerically, we approximate the second-order time derivative in equation (4.120) by the finite-difference expression (4.41). This leads for equation (4.120) to

$$x_{i-1} - 2x_i + x_{i+1} = \tilde{f}_i, \quad x_0 = 0, \quad x_N = b, \qquad (4.121)$$

where $\tilde{f}_i \equiv F_i \Delta t^2/m$ and $F_i \equiv F(t_i)$. As can be seen from equation (4.121), the second-order differential equation is to be evaluated at three successive positions along the discretized $x$-axis, i.e., $x_{i-1} = x(t_{i-1})$, $x_i = x(t_i)$, and $x_{i+1} = x(t_{i+1})$. The discretized times are given by $t_i \equiv i\Delta t$, where $i = 1, 2, \ldots, N - 1$. The time segment $\Delta t$ is defined as $\Delta t \equiv t_f/N$, with $N$ denoting the total number of time segments of size $\Delta t$. Finally, the boundary values at the two end points require that $x_0 = 0$ and $x_N = b$, as shown in equation (4.121). Equation (4.121) constitutes a system of $N - 1$ coupled linear equations,

$$
\begin{aligned}
i = 1: \quad & x_0 - 2x_1 + x_2 = \tilde{f}_1, \\
i = 2: \quad & x_1 - 2x_2 + x_3 = \tilde{f}_2, \\
i = 3: \quad & x_2 - 2x_3 + x_4 = \tilde{f}_3, \\
& \cdots \qquad\qquad \cdots \\
i = N - 2: \quad & x_{N-3} - 2x_{N-2} + x_{N-1} = \tilde{f}_{N-2}, \\
i = N - 1: \quad & x_{N-2} - 2x_{N-1} + x_N = \tilde{f}_{N-1},
\end{aligned}
\tag{4.122}
$$

which determine the $N - 1$ unknown positions $x_1, x_2, \ldots, x_{N-1}$. Written in matrix form, this system of equations if given by

$$
\begin{pmatrix}
-2 & 1 & & & & \\
1 & -2 & 1 & & & \\
 & 1 & -2 & 1 & & \\
 & & \ddots & \ddots & \ddots & \\
 & & & 1 & -2 & 1 \\
 & & & & 1 & -2
\end{pmatrix}
\begin{pmatrix}
x_1 \\ x_2 \\ x_3 \\ \vdots \\ x_{N-2} \\ x_{N-1}
\end{pmatrix}
=
\begin{pmatrix}
\tilde{f}_1 - x_0 \\ \tilde{f}_2 \\ \tilde{f}_3 \\ \vdots \\ \tilde{f}_{N-2} \\ \tilde{f}_{N-1} - x_N
\end{pmatrix},
\tag{4.123}
$$

where the coefficient matrix is tridiagonal and symmetric. Such systems can be solved numerically with standard numerical routines, such as `tridag` from the Numerical Recipes [2].

# References

[1] Heath M T 2002 *Scientific Computing: An Introductory Survey* 2nd edn (New York: McGraw-Hill)

[2] Press W H, Teukolsky S A, Vetterling W T and Flannery B P 2007 *Numerical Recipes: The Art of Scientific Computing* 3rd edn (Cambridge: Cambridge University Press)

[3] Anderson E *et al* 1999 *LAPACK Users' Guide* 3rd edn (Philadelphia, PA: Society for Industrial and Applied Mathematics (SIAM))

[4] Dongarra J, Moler C, Bunch J and Stewart G 1979 *LINPACK Users* (Philadelphia, PA: Society for Industrial and Applied Mathematics (SIAM))

[5] IMSL. IMSL Fortran Numerical Library Documentation available from IMSL distributors, e.g., Rogue Wave Software

[6] The Numerical algorithms Group 2024 NAG Fortran Library Available from https://www.nag.com/

[7] de Boor C 1978 *A Practical Guide to Splines* (New York: Springer)
[8] Schumaker L L 2007 *Spline Functions: Basic Theory* (Cambridge: Cambridge University Press)
[9] Fornberg B 1988 Generation of finite difference formulas on arbitrarily spaced grids *Math. Comput.* **51** 699–706
[10] Saad Y 2003 *Iterative Methods for Sparse Linear Systems* (Philadelphia, PA: Society for Industrial and Applied Mathematics (SIAM))
[11] Young D M 1971 *Iterative Solution of Large Linear Systems* (New York: Academic)

**IOP** Publishing

# Introduction to Computational Physics for Undergraduates (Second Edition)

**Omair Zubairi and Fridolin Weber**

# Chapter 5

## Problem solving methodologies

Computational physics is a subject where computing is used to gain insight into complex systems. It is highly multidisciplinary involving physics, mathematics, and computer science, as illustrated in figure 5.1.

This requires knowledge in the Linux/Unix environment, a programming language of some sort, and numerical techniques. Having this knowledge from the previous chapters, we can now attack a variety of problems in science and engineering that will require computing of some sort. In this chapter, we layout a general guideline on methods and strategies that can be applied to any type of computational problem.

When solving various problems in physics (and other scientific and engineering disciplines), we have to first ask ourselves a few questions:
- What is the physics problem at hand?
- What is the mathematical model which describes this particular problem?
  - What does the model tell us about the physics problem?
  - Is it realistic?
  - Can a solution be obtained in a reasonable amount of time and effort?
- Which programming language would work best for the application?

Asking the above questions is the first step to solving complex computational types of problems. Once you have determined the answers to these questions, you are ready to apply your skills in computing to solve the problem at hand.

## 5.1 General guidelines

A general guideline to solving computational physics problems is as follows:
1. Analyze the problem and think about it rationally.
2. Plan out the program, write out what you think you need to do.

doi:10.1088/978-0-7503-6493-5ch5

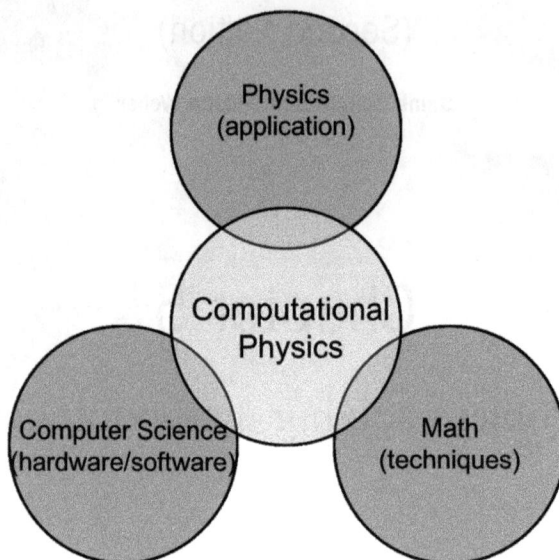

**Figure 5.1.** Venn diagram illustrating a multidisciplinary approach to computational physics.

3. Draw a flowchart explaining the features of your program.
4. Write pseudo-code to match your flow diagram.
5. Write your source code.
6. Compile your source code.
7. Execute (run) your source code (debug if necessary).

The seven steps listed here are very useful and can be applied not only to physics problems but also many other problems in science and engineering. A great example of applying these guidelines is on the topic of two-dimensional kinematics, particularly on projectile motion. Projectile motion is a topic encountered by many students in a first semester physics course and can be applied to a variety of other applications.

## 5.2 Projectile motion example

Consider an object which is projected upward with some initial velocity $v_0$ at some given angle $\theta$. For this scenario, we can ask the following questions:
- What is the maximum height the object reaches?
- How long will the object be in the air before it hits the ground?
- How far away is the object when it hits the ground?

To answer these questions, we can look at the trajectory of this object and computationally compute this by applying the seven steps listed above.

**1. Analyze the problem and think about it rationally.**

To solve this problem, we would need mathematical expressions which will describe the $x$ and $y$ positions. This would require us to use the well-known kinematic equations, as described in the expressions in equations (5.1)–(5.3).

$$v_x = v_{x0} + a_x t \qquad\qquad v_y = v_{y0} + a_y t, \tag{5.1}$$

$$x = x_0 + v_{x0} t + \frac{1}{2} a_x t^2 \qquad\qquad y = y_0 + v_{y0} t + \frac{1}{2} a_y t^2, \tag{5.2}$$

$$v_x^2 = v_{x0}^2 + 2a_x(x - x_0) \qquad\qquad v_y^2 = v_{y0}^2 + 2a_y(y - y_0). \tag{5.3}$$

Since we are looking for the $x$ and $y$ positions, we will utilize the expressions

$$x = x_0 + v_{x0} t + \frac{1}{2} a_x t^2, \qquad\qquad y = y_0 + v_{y0} t + \frac{1}{2} a_y t^2, \tag{5.4}$$

where, if we assume that the initial positions in the $x$ and $y$ direction are zero, in addition to knowing that the horizontal acceleration $a_x$ is zero and the vertical acceleration $a_y$ is simply the acceleration due to gravity, our expressions which describe the $x$ and $y$ positions modify to

$$x = v_0 \cos(\theta) t \tag{5.5}$$

$$y = v_0 \sin(\theta) t - \frac{1}{2} g t^2 \tag{5.6}$$

where we have substituted the expressions for $v_{0x}$ and $v_{0y}$ with $v_0 \cos(\theta)$ and $v_0 \sin(\theta)$, respectively. This is an important concept because velocity is a vector and must be broken up into its components.

**2. Plan out the program.**

Having our expressions which will describe the $x$ and $y$ positions, we can now start to plan out our program. As the projectile moves, the horizontal and vertical positions will change over time, which will require the program to *update* these positions. Thus,

- Updating the positions would require some sort of *iterative* process (i.e. loops).
- Would also need some sort of *step-size* (i.e. a time step).
- Would need to *write* this data out as it is being calculated.
- Would also need to update the step-size for each *new* position.
- Finally, we would need iterate this *until* the projectile hits the ground (i.e. requires a logical condition).

**3. Draw out a flowchart.**

A typical flowchart for this type of problem is illustrated in figure 5.2. The flowchart describes the important points of the program, such as input parameters, iterative processes for our positions, writing data, and logical conditions.

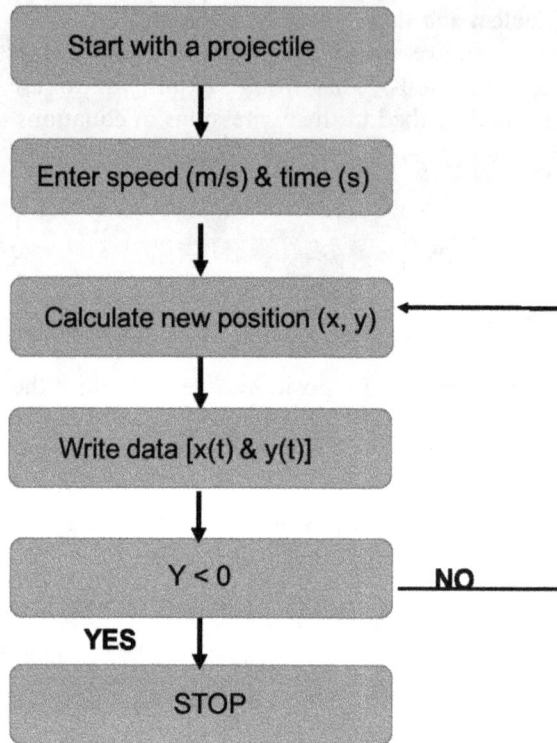

**Figure 5.2.** Flow chart illustrating calculations of projectile motion.

### 4. Write out pseudo-code.

We can easily write some simple pseudo-code to describe the main processes happening within our program. The pseudo-code below describes what will be happening within our iterative process.

```
Do i = start, finish
        time =···
        x=···
        y=···
        Write out data
        Update time-step
        Iterate UNTIL Projectile Hits the Ground
End Do
```

**5. Write your source code.**

Using the pseudo-code, one can easily see that the major part of the program is within the loop. The actual code is left to the student to exercise their programming skills; however, the main loop is provided for your reference:

```
Do i = 0,1000
        t = i*dt

        x = v0*cos(theta)*t
        y = v0*sin(theta)*t - 0.5*g*t**2

        t = t + dt
        Write(15,*) x, y
        If (y < 0) Exit
End Do
```

Once you have successfully written the entire source code, you can compile and run the program, and then graphically illustrate the trajectory of an object projected upward with some initial velocity and given angle.

Figure 5.3 shows the trajectories for a projectile launched with various velocities and angles by computing the expressions in equation (5.4) via the main loop listed above.

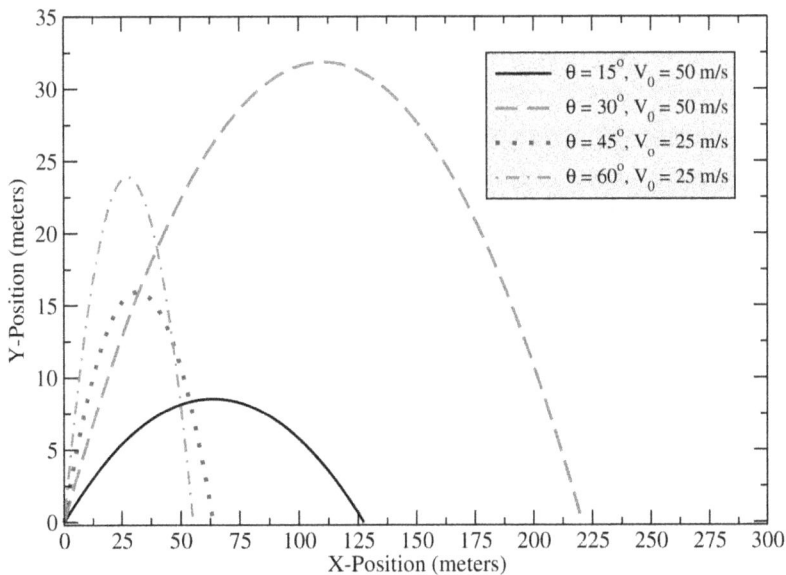

**Figure 5.3.** $x$ and $y$ positions of an object projected upward with various initial velocities and angles.

By applying our steps outlined in this chapter, we can attack a variety of physics problems. The worksheets and homework assignments listed in the next two chapters will give ample opportunities to apply these general guidelines.

**IOP** Publishing

# Introduction to Computational Physics for Undergraduates (Second Edition)

**Omair Zubairi and Fridolin Weber**

# Chapter 6

## Worksheet assignments

## 6.1 Coding a mathematical expression

*Purpose: Compute the value of a given mathematical expression.*

Given is the following mathematical expression:

$$\chi = (\pi + \sin(x)\cos^3(x)\,\sqrt{1 + \sqrt{x}} + e^{-x\sin(x)})(1 + x^2 + x^3 + x^4)^{-2} + \pi^2.$$

**Tasks:**

Write a structured Fortran 90 program which computes the function above for a given value of $x$.

**Program design:**

1. The variable $x$ is keyboard input.
2. Make use of the Fortran 90 intrinsic functions `SIN(X)`, `COS(X)`, `SQRT(X)`, and `EXP(X)`.
3. Use the `Parameter` statement for $\pi$.
4. The value of $\chi$ is terminal output.
5. Run your code for $x = 0, 0.5, 1.0, 1.5$ and compare the results for $\chi$ with the results obtained by your fellow students.

A sample code for this problem is shown below. A detailed description of the different program steps and good coding practices can be found in Appendix D.

doi:10.1088/978-0-7503-6493-5ch6     6-1

```
program evaluate_function
  implicit none
  real :: x, result

  real, parameter :: pi = ACOS(-1.0)

  ! Prompt user for input values
  write(*,*) 'Enter a value for x (e.g., 0, 0.5, 1.0, 1.5):'
  read(*, *) x

  ! Evaluate the function
  result = (pi + SIN(x) * COS(x)**3 * SQRT(1.0 + SQRT(x)) + EXP(-x&
        & * SIN(x))) / (1.0 + x**2 + x**3 + x**4)**2 + pi**2

  ! Print the result
  print *, 'chi(', x, ') = ', result
end program evaluate_function
```

The screen output generated by this program for $x = 1.0$ is

```
chi(1.0000000) = 10.104 628 6
```

## 6.2 Comparing two functions

*Purpose: Practice the use of DO loops and the OPEN statement and generate graphical output.*

Given are the following two functions,

$$\phi(x) = e^{-x^2 \sin(x)^2} x^{3/2},$$

$$\tau(x) = 0.124\,523 + 0.739\,594\,(x - 0.25) + 0.657\,81\,(x - 0.25)^2 - 0.916\,955$$
$$(x - 0.25)^3 - 0.214\,698\,(x - 0.25)^4 - 2.351\,54\,(x - 0.25)^5,$$

where $\tau(x)$ is the Taylor expansion of $\phi(x)$ for $x \in [0, 1]$.

**Tasks:**

Write a structured Fortran 90 program which computes $\phi(x)$ and $\tau(x)$ for a range of $x$ values.

**Program design:**

1. Use the range $x = 0, 1$ with a step size of 0.001.
2. Using the OPEN statement, design your code such that the results for $\phi(x)$ and $\tau(x)$ are written to two different output (data) files.
3. Declare the purpose of your code and all Fortran variables in the preamble of the program.
4. Produce a plot which shows $\phi(x)$ and $\tau(x)$ for $0 \leqslant x \leqslant 1$.

**Sample code:**

```
program comparison
   implicit none
   real :: x, s, Phi_x, tau_x, p1, p2, p3, p4, p5, p6
   real :: a=0.0, b=1.0, Delta=0.001
   integer :: N, i

   p1=0.124523; p2=0.739594; p3=0.65781; p4=0.916955; p5=0.214699;&
      & p6=2.35154

   N=INT((b-a)/Delta)

   open(unit=100, file='Phi.dat', status='unknown')
   open(unit=200, file='tau.dat', status='unknown')

Loop: do i=0,N
         x = a + Delta * i
      Phi_x = EXP( -x**2 * SIN(x)**2 ) * x**(3.0/2.0)
      s     = x - 0.25
      tau_x = p1 + p2*s + p3 * s**2 - p4 * s**3 - p5 * s**4 - p6 * &
         &s**5
         write(100,*) x, phi_x
         write(200,*) x, tau_x
      end do Loop
   close(unit=100); close(unit=200)
   end program comparison
```

The data produced by this program for $\phi(x)$ and $\tau(x)$ over the range $0 \leqslant x \leqslant 1$ are shown graphically in figure 6.1.

## 6.3 Bessel functions of the first kind

*Purpose: Practice the DO loop and OPEN constructs and generate graphical output.*

Bessel functions are used in optics to characterize the pattern observed when light is focused by a perfect lens with a circular aperture. The $\alpha$th-order Bessel functions of the first kind, denoted as $J_\alpha(x)$, are solutions to Bessel's differential equation that are finite at the origin ($x = 0$) for integer or positive $\alpha$, and diverge as $x$ approaches zero for negative non-integer $\alpha$ values. The function can be defined by its Taylor series expansion around $x = 0$,

$$J_\alpha(x) = \sum_{p=0}^{\infty} \frac{(-1)^p}{p! \, \Gamma(p + \alpha + 1)} \left(\frac{x}{2}\right)^{2p+\alpha}, \tag{6.1}$$

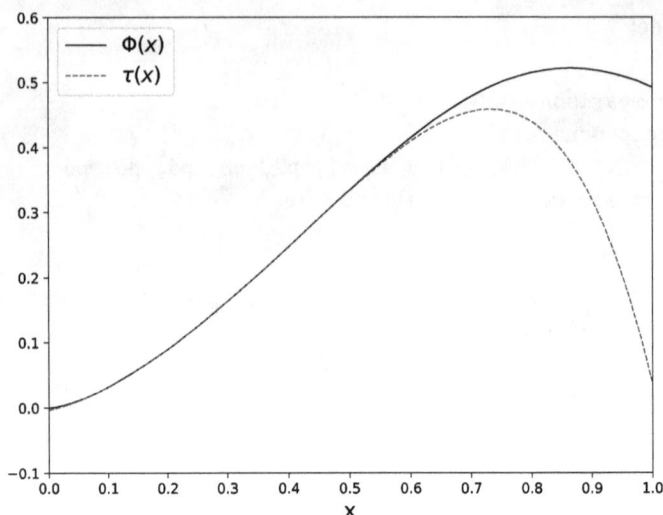

**Figure 6.1.** Graphical illustration of the functions $\phi(x)$ and $\tau(x)$.

where $\Gamma(n) = (n - 1)!$ $(n > 0)$ is the gamma function. The zero-order Bessel function, $J_0(x)$, follows from (6.1) as

$$J_0(x) \approx \sum_{p=0}^{m} \frac{(-1)^p}{p!\,\Gamma(p + 1)}\left(\frac{x}{2}\right)^{2p}. \tag{6.2}$$

**Tasks:**

Write a structured Fortran program that computes $J_0(x)$ from equation (6.2) for $m = 30$, where $x$ ranges from $\varepsilon$ to 10, with $\varepsilon = 0.1$. The program should utilize double precision for accurate floating-point calculations. This can be achieved using the Fortran syntax:

```
IMPLICIT NONE
INTEGER, PARAMETER :: DP = SELECTED_REAL_KIND(15)
  REAL(KIND = DP) :: ...    ! list of your variables
```

**Program design:**

1. Utilize a DO loop construct to compute $J_0(x)$ over the interval $\varepsilon \leqslant x \leqslant 10$ with a step size $\Delta x = 0.1$. Use the OPEN statement to write the results of $J_0(x)$ to an external file.
2. Demonstrate the impact of varying the number of terms $m$ when calculating $J_0(x)$. Compute $J_0(x)$ for $\varepsilon \leqslant x \leqslant 10$ using $m = 5$, 10, 15, and 20.
3. Illustrate the results of steps 1 and 2 graphically.
4. Extend the program to compute $J_\alpha(x)$ from equation (6.1) for a given $\alpha$. Compute $J_\alpha(x)$ for $\varepsilon \leqslant x \leqslant 10$ using $m = 20$, and write the results to an external file. Run your code for $\alpha = 0$, 1, 2, 3, 4, and 5 (with $m = 20$ in each case), and compare the results graphically on a single plot.

**Sample code:**

```
PROGRAM Bessel_J0
    IMPLICIT NONE
    INTEGER, PARAMETER :: DP = SELECTED_REAL_KIND(15) ! Double
                                                       ! precision
    REAL(KIND=DP) :: x, sum, numerator, denominator
    INTEGER :: m=20, p, i
    REAL, PARAMETER :: delta = 0.1_DP ! Step size for x values
    INTEGER, PARAMETER :: n_x = 100   ! Number of x values (0 to 10
                                       ! with delta=0.1)


    OPEN(unit=10, file='J_0.dat', status='unknown')

    WRITE(*,*) "Computing J_0(x) for m = 20 and x values from &
            &[0.1,10]"

 x_values:   DO i = 1, n_x
                x = REAL(i) * delta

                sum = 0.0_DP ! Initialize sum for J_0(x)

        sum_p: DO p = 0, m
                numerator   = (-1)**p * (x/2.0_DP)**(2*p)
                denominator = GAMMA(REAL(p+1)) * GAMMA(REAL(p+1))
                sum = sum + numerator / denominator
        END DO sum_p

        WRITE(10, '(2F12.8)') x, sum

    END DO x_values

        WRITE(*,*) "Zero-order Bessel function J_0(x) computed and saved&
                & to 'J_0.dat'"

        CLOSE(unit=10)
    END PROGRAM Bessel_J0
```

## 6.4 Logical IF statements

*Purpose: Practice logical IF statements and use Fortran intrinsic functions.*
The following function $\Phi(x, y, z)$ is given for various conditions:

$$\Phi(x, y, z) = \begin{cases} \sqrt{x^3 + y^3 + z^3} & \text{if } x < 0 \\ \pi/4 & \text{if } x = 0, \\ \sin(xy) + \cos(xz) & \text{if } x > 0 \end{cases} \quad (6.3)$$

**Tasks:**

Write a structured Fortran program that reads in three real numbers $(x, y, z)$ and computes $\Phi(x, y, z)$ for the three conditions listed above.

**Program design:**

1. Comment the different steps in your program.
2. Make use of the logical IF statement to discriminate between $x < 0$, $x = 0$, and $x > 0$
3. Make use of the PARAMETER statement for $\pi$.
4. Run your program for the following values: $(x, y, z) = (-1.5, 4.0, 9.0)$, $(0.0, 12.0, -2.2)$, and $(3.5, 4.0, 0.5)$. Utilize the SELECT CASE construct to evaluate the three cases.

**Sample code:**

```
PROGRAM PhiFunction
  IMPLICIT NONE
  REAL :: x, y, z, phi
  REAL, PARAMETER :: pi = ACOS(-1.0)
  INTEGER :: i

  ! Run the program for specific x,y,z values
  select: DO i = 1, 3

          SELECT CASE (i)
                CASE (1)
                x = -1.5; y =  4.0; z =  9.0
                CASE (2)
                x =  0.0; y = 12.0; z = -2.2
                CASE (3)
                x = 3.5; y = 4.0; z = 0.5
          END SELECT

  ! Compute Phi(x, y, z) based on conditions
  x_value: IF (x < 0) THEN
              phi = SQRT(x**3 + y**3 + z**3)
            ELSE IF (x == 0) THEN
              phi = pi / 4.0
            ELSE
              phi = SIN(x * y) + COS(x * z)
            END IF x_value

  ! Display the result
      WRITE(*,*) 'Phi(', x, ',', y, ',', z, ') = ', phi
   END DO select
END PROGRAM PhiFunction
```

The results are given by

```
Phi(-1.5000000,  4.000 000 0,  9.000 000 0) = 28.100 267 4
Phi(0.000 000 0, 12.000 000 0, -2.2000000) = 0.785 398 2
Phi(3.500 000 0,  4.000 000 0,  0.500 000 0) = 0.812 361 4
```

## 6.5 Lead concentration in humans

*Purpose: Practice DO loops, logical IF statements, and Input/Output data handling.*

Lead is widely present in our environment due to its natural occurrence and human activities that have introduced it into the general environment. Because lead may be present in environments where food crops are grown and animals used for food are raised, various foods may contain unavoidable but small amounts of lead that do not pose a significant risk to human health. Small amounts of lead in adults are not thought to be harmful. However, even low levels of lead can be very dangerous to infants and children. According to the Centers for Disease Control and Prevention (CDC), blood lead levels of 2.4 $\mu$mole/10 L (5 $\mu$g/dL) or greater require further testing and monitoring in children[1].

For this worksheet, you will write a Fortran program which reads lead concentration data measured in children from a data file. For this data set, the program will then calculate: the mean $\langle x \rangle$,

$$\langle x \rangle = \frac{1}{n} \sum_{i=1}^{n} x_i, \tag{6.4}$$

the variance $\sigma^2$,

$$\sigma^2 = \frac{1}{n-1} \sum_{i=1}^{n} (x_i - \langle x \rangle)^2, \tag{6.5}$$

and the standard deviation $\sigma$,

$$\sigma = \sqrt{\frac{1}{n-1} \sum_{i=1}^{n} (x_i - \langle x \rangle)^2}, \tag{6.6}$$

where $n$ is the number of measured data (observations). Recall that $\sigma$ is a very useful measure of the scatter of observed data. A range extending one $\sigma$ above and below the mean $\langle x \rangle$ includes about 68% of the observations. Extending this range to two $\sigma$ above and below the mean captures about 95% of the observations, and extending it

---

[1] 5 $\mu$g/dL stands for 5 micrograms per deciliter (dL).

to three $\sigma$ includes about 99.7% of the observations. Thus, by considering one, two, or three standard deviations above and below the mean, one can estimate the ranges expected to encompass approximately 68%, 95%, and 99.7% of the observed data, respectively.

The probability distribution (function)

$$f(x \mid \langle x \rangle, \sigma^2) = \frac{1}{2\sigma^2\pi} \, e^{-(x-\langle x \rangle)^2/2\sigma^2} \tag{6.7}$$

of a set of data is represented by a curve defined uniquely by two parameters, which are the mean and the standard deviation of a given data set. The curve is always symmetrically bell shaped, but the extent to which the bell is compressed or flattened out depends on the standard deviation of a given data set.

**Tasks:**

Write a structured Fortran program that reads lead concentrations (in $\mu$mole/10 L) measured in children from an external data file and calculates the mean value, variance, and standard deviation, as defined in equations (6.4)–(6.6). The data for this set is given by the following array:

```
[0.1,0.4,0.6,0.8,1.1,1.2,1.3,1.5,1.7,1.9,1.9,2.0,2.2,2.6,3-
.2]
```

You will have to manually create an external file (.dat) for the data listed above which then you will READ into your program.

**Program design:**

1. The screen output generated by the program should be as follows:

```
The Mean of this data set is:
The variance of this data set is:
The standard deviation is:
The number of data points is:
```

2. Make use the implicit DO loop construct (i.e., READ(unitnumber, *, end=unitnumber)) to read the data from the data file.
3. If the number of input data is less than 2, the program should tell (screen output) the user that the number of input data is insufficient to carry out a statistical analysis. Use the CALL EXIT statement to terminate the program.
4. The probability function $f(x \mid \langle x \rangle, \sigma^2)$ is to be computed and graphically illustrated for $-2 \ \mu$mole/10L $\leqslant x \leqslant 5 \ \mu$mole/10L. To cover this range, make use of $x_k = a + (b - a)k/N$, with $N = 100$.
5. In your plot, mark the locations where 68% ($\langle x \rangle \pm \sigma$), 95% ($<x> \pm 2\sigma$), and 99.7% $<\langle x \rangle > \pm 3\sigma$ of the data are located.

## Sample code:

```fortran
PROGRAM ProbabilityDensity
    IMPLICIT NONE
    INTEGER :: i, n = 1, noGridPoints, no_of_data
    REAL :: std_dev, sum, variance, Mean, x_i, f_i, a, b
    REAL :: x(100)
    REAL, PARAMETER :: pi = ACOS(-1.0)

    OPEN(unit=100, file='LeadCon.dat', status='unknown')
    OPEN(unit=101, file='ProbDensity.dat', status='unknown')
    ! Read data from file
    DO
        READ(100, *, end=200) x(n)
        n = n + 1
    END DO
200 CONTINUE
    no_of_data = n - 1 ! Number of input data points

    ! Compute the mean value
    sum = 0.0
    DO i = 1, no_of_data
        sum = sum + x(i)
    END DO
    Mean = sum / REAL(no_of_data)

    ! Compute the variance
    sum = 0.0
    DO i = 1, no_of_data
        sum = sum + (x(i) - Mean)**2
    END DO
    variance = sum / REAL(no_of_data - 1)
    std_dev = SQRT(variance) ! Compute standard deviation

    ! Output results
    WRITE(*, *) 'The Mean of this data set is:', Mean
    WRITE(*, *) 'The variance of this data set is:', variance
    WRITE(*, *) 'The standard deviation is:', std_dev
    WRITE(*, *) 'The number of data points is:', no_of_data

    ! Data for plotting the probability density function
    a = -2.0
    b =  5.0
    noGridPoints = 100
    DO i = 1, noGridPoints
        x_i = a + (b - a) * REAL(i) / REAL(noGridPoints)
        f_i = EXP(-(x_i - Mean)**2 / (2.0 * variance)) / SQRT(2.0 *&
            & variance * pi)
        WRITE(101, *) x_i, f_i
    END DO
END PROGRAM ProbabilityDensity
```

The output produced by this code includes:

```
The mean of this data set is: 1.500 000 0
The variance of this data set is: 0.711 428 6
The standard deviation is: 0.843 462 3
The number of data points is: 15
```

## 6.6 Nested Do loops and double summations

*Purpose: Utilize the nested DO loop construct to compute double and triple sums, as well as the product of a sequence of term.*

**Tasks:**

Write a structured Fortran program that computes and prints out the results of the following expressions for $N = 100\ (100)\ 500$:

$$A(N) = \left(120 \sum_{i=1}^{N}\sum_{j=1}^{i} j^{-2}\ (i+1)^{-2}\right)^{1/4}, \qquad (6.8)$$

$$B(N) \equiv 2 \prod_{k=1}^{N-1} \frac{(2k)^2}{(2k-1)(2k+1)}, \qquad (6.9)$$

$$C(N) = \left(32\left(\sum_{k=0}^{N} \frac{k+1}{(2k+1)^2\ (2k+3)} - \frac{1}{8}\right)\right)^{1/2}, \qquad (6.10)$$

$$D(N) = \left(216\sum_{i=1}^{N}\sum_{j=1}^{N}\sum_{k=1}^{N} \frac{1}{(i\,j\,k)^2}\right)^{1/6}. \qquad (6.11)$$

The results for $A(N)$, $B(N)$, $C(N)$, and $D(N)$ are to be compared with the analytic results given by $A(\infty) = \pi = 3.141\ 593$, $B(\infty) = \pi$, $C(\infty) = \pi$, and $D(\infty) = \pi$. The symbol $\prod$ in equation (6.9) denotes the $\prod$-product. It is similar to the summation symbol $\sum$, but instead of addition its operation is multiplication. For example, $\prod_{k=1}^{N} f_k = f_1 \cdot f_2 \cdot f_3 \cdot \ldots f_N$.

**Program design:**

1. There is no keyboard input.
2. Use pi = ACOS(-1.0) in a PARAMETER statement to define $\pi$.
3. Utilize an outer DO loop to determine the value of $N$. The values of $N$ run from 100 to 500 in steps of 100.
4. Inside this outer loop, include nested DO loop constructs to compute the expressions shown in equations (6.8) through (6.11).
5. All DO loops must be named.

6. For each $N$ value, the results for $A(N)$, $B(N)$, $C(N)$, and $D(N)$ are to be written to standard output together with the relative deviation $\Delta$ of these values from $\pi$, i.e., $\Delta = |(A(N) - \pi)|/\pi$, and similarly for $B(N)$, $C(N)$, and $D(N)$.

The terminal output produced by the code should be as shown below:

```
N = 100
A = 3.125 841 62  (A_N-pi)/pi  = 5.013 738 29E-03
B = 3.133 709 19  (B_N-pi)/pi  = 2.509 411 42E-03
C = 3.135 298 73  (C_N-pi)/pi  = 2.003 446 22E-03
D = 3.132 076 74  (D_N-pi)/pi  = 3.029 037 03E-03
   ... ... ...

   ... ... ...

   ... ... ...
N = 500
A = 3.138 435 36  (A_N-pi)/pi  = 1.005 024 42E-03
B = 3.140 020 61  (B_N-pi)/pi  = 5.004 251 73E-04
C = 3.140 322 45  (C_N-pi)/pi  = 4.043 472 00E-04
D = 3.139 684 20  (D_N-pi)/pi  = 6.075 073 62E-04
```

## 6.7 Least squares fit

*Purpose: Fit a set of experimental data with a mathematical model (using the least-squares technique).*

A satellite experiment was launched in the NIMBUS 7 spacecraft in 1978 to collect data on the composition and structure of the middle atmosphere. The instrumentation and sensors collected data from October 25, 1978 to May 28, 1979, returning more than 7000 sets of data to the Earth each day. These data were used to determine temperature, ozone, water, vapor, nitric acid, and nitrogen dioxide distributions in the stratosphere and mesosphere. (The stratosphere and the mesosphere are atmospheric layers around the Earth from about 15 km to approximately 85 km above the Earth's surface.) Assume that we have collected a set of data measuring the ozone mixing ratios in parts per million volume (ppmv), as shown in table 6.1. Over small regions, these data are nearly linear, and thus we can use a linear model to estimate the ozone at altitudes other than the ones for which we have specific data.

**Tasks:**

Write a structured Fortran program that reads a data file generated by the data set in table 6.1 and performs a linear fit on the data, following the methods outlined in chapter 5, specifically detailed in section 4.1.1. Utilize the least-squares technique to compute and output the parameters of the best-fit linear model along with the average squared error.

**Table 6.1.** Atmospheric data measuring ozone mixing ratios (OMR) [1]

| Altitude (km) | OMR (ppmv) | Altitude | OMR (ppmv) |
|---|---|---|---|
| 20.0 | 3.1 | 34.0 | 9.3 |
| 22.0 | 4.01 | 35.0 | 7.7 |
| 22.4 | 4.1 | 35.3 | 7.8 |
| 23.0 | 4.3 | 36.0 | 7.7 |
| 24.0 | 4.4 | 37.0 | 8.1 |
| 26.0 | 5.3 | 38.0 | 8.4 |
| 28.0 | 6.1 | 39.4 | 9.0 |
| 29.7 | 7.2 | 40.0 | 8.9 |
| 31.0 | 8.3 | 41.0 | 9.6 |
| 33.0 | 7.5 | 41.5 | 10.7 |

**Program design:**

1. The original data and the best-fit model data are to be written to an output file.
2. The screen output produced by your program should be as follows:

```
Linear model: y = -2.3955 + 0.2977 x
original original best-fit residual
x y Y r
20.00 3.10 3.56 -0.46
22.00 4.01 4.15 -0.14
22.40 4.10 4.27 -0.17
39.40 9.00 9.33 -0.33
... ... ... ...
... ... ... ...
... ... ... ...
40.00 8.90 9.51 -0.61
41.00 9.60 9.81 -0.21
41.50 10.70 9.96 0.74
Averaged squared error (chi) = 0.3921
```

3. Generate a plot which has the original data and the best-fit (least-squares) linear model in the same graph.

A graphical illustration of the original data and the linear best-fit is depicted in figure 6.2.

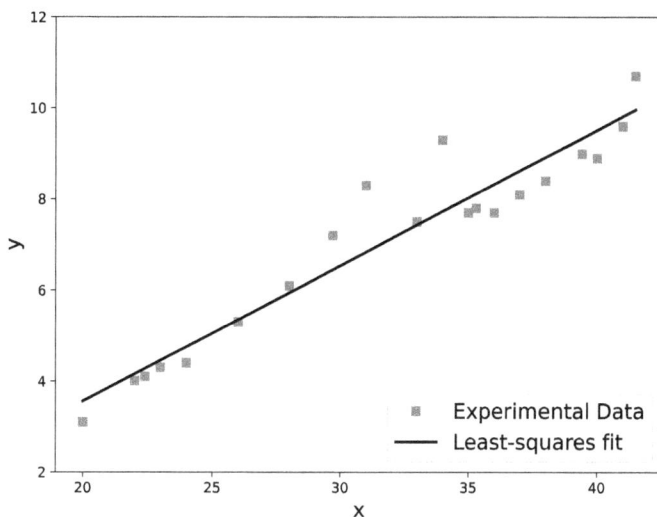

**Figure 6.2.** Theoretical fit of a set of experimental data.

## 6.8 Numerical derivatives

*Purpose: Illustrate how to numerically approximate derivatives.*

The first derivative of a function can be approximated via the forward Euler's method and the three-point rule formula, as described in chapter 4. While, these methods serve their purpose for approximating the first derivative, higher-order methods are required for higher-order derivatives of functions.

For second-order derivatives, one can use the three-point rule given by equation (4.41) and the five-point rule given by equation (4.42).

**Tasks:**

Write a well-commented structured Fortran program which numerically approximates the second derivative for the function:

$$f(x) = x \sin(x) \qquad (6.12)$$

at the point of $x_0 = 26.0$.

**Program design:**
1. Declare all variables with quad precision i.e. Real (kind =16)::
2. Choose a step size such that dx = 0.1
3. Use a function to compute the exact second-order derivative of $f(x) = x \sin(x)$.
4. Calculate the error between the exact solution and the approximations.
5. Comment on your graph, what can you say about the slope for each?

## Sample code:

```fortran
program second_derivative_comparison
  implicit none
  real(kind=16) :: x0, dx, exact_second_derivative
  real(kind=16) :: f_x0, f_x0_pdx, f_x0_mdx
  real(kind=16) :: f_x0_p2dx, f_x0_m2dx
  real(kind=16) :: approx_second_derivative_3point,&
                 & approx_second_derivative_5point
  real(kind=16) :: error_3point, error_5point

  x0 = 26.0; dx = 0.1

! Exact second derivative of the function f(x) = x*sin(x)
  exact_second_derivative = second_derivative_exact(x0)

! Calculate the function values needed for the approximations
  f_x0     = x0 * sin(x0)
  f_x0_pdx = (x0 + dx) * sin(x0 + dx)
  f_x0_mdx = (x0 - dx) * sin(x0 - dx)
  f_x0_p2dx = (x0 + 2.0 * dx) * sin(x0 + 2.0 * dx)
  f_x0_m2dx = (x0 - 2.0 * dx) * sin(x0 - 2.0 * dx)
! Three-point central difference approximation for the second
! derivative
  approx_second_derivative_3point = (f_x0_pdx - 2.0 * f_x0 +&
                              & f_x0_mdx) / (dx * dx)

! Five-point central difference approximation for the second
! derivative
  approx_second_derivative_5point = (-f_x0_p2dx + 16.0 * f_x0_pdx -&
      & 30.0 * f_x0 + 16.0 * f_x0_mdx - f_x0_m2dx) / (12.0 * dx * dx)

! Calculate the errors between the exact and approximate second
! derivatives
  error_3point = abs(approx_second_derivative_3point -&
            & exact_second_derivative)
  error_5point = abs(approx_second_derivative_5point -&
            & exact_second_derivative)

! Output the results and errors to the screen
  print *, 'Exact second derivative:           ', exact_second_derivative
  print *, 'Three-point rule approximation: ', approx_second_derivative_3point
  print *, 'Error using three-point rule:   ', error_3point
  print *, 'Five-point rule approximation:  ', approx_second_derivative_5point
  print *, 'Error using five-point rule:    ', error_5point

contains

! Function to compute the exact second derivative of f(x) = x*sin(x)
  real(kind=16) function second_derivative_exact(x)
    real(kind=16), intent(in) :: x
    second_derivative_exact = 2.0 * sin(x) - x * sin(x)
  end function second_derivative_exact

end program second_derivative_comparison
```

The program yields the following results:

```
Exact second derivative: -18.30
Three-point rule approximation: -18.77
Error using three-point rule: 0.4688
Five-point rule approximation: -18.89
Error using five-point rule: 0.5938
```

## 6.9 Numerical integration

*Purpose:* You will learn how to integrate a continuous function numerically using the Trapezoidal rule.

**Tasks:**

Write a structured and well-commented Fortran program that uses the Trapezoidal rule given by equation (4.50) to compute the integral

$$\int_a^b dx\, x^{5/2} \cosh^2(x) e^{-\sqrt{x}\,\cos(x) - x^3 \sin^2(x^2)} (1.0 + 0.5x + 0.2x^2 + 0.1x^4)^{-5} \quad (6.13)$$

for integration limits $(a = 0, b = 0.5)$, $(a = 0, b = 1.0)$, and $(a = 1.0, b = 1.6)$. The number of grid points, $N$, to compute the integral of equation (6.13) should range from 2 to 10, in increments of 2 (figure 6.3).

**Program design:**

1. All variables are declared with double precision using the dp kind parameter as follows: `integer, parameter :: dp = selected_real_kind (15, 307)`

2. The program prompts the user to enter the integration limits $a$ and $b$.

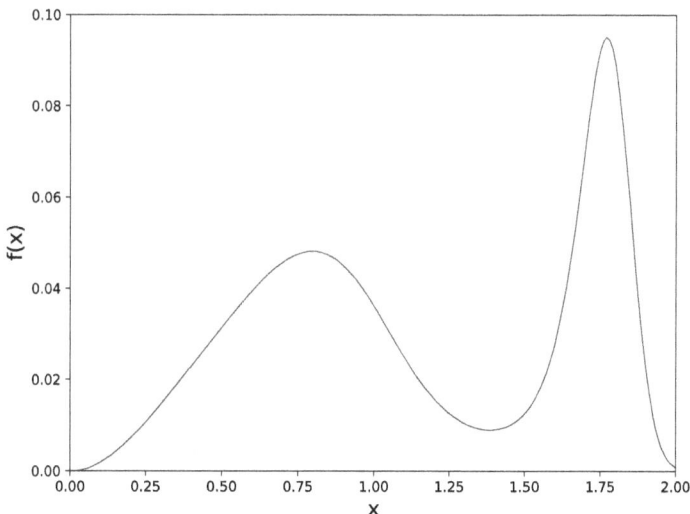

**Figure 6.3.** Graphical illustration of the integrand of equation (6.13).

3. The program loops over the number of grid points $N$ from 2 to 100 in steps of 2.
4. The step size $h$ is calculated as $(b - a)/N$.
5. The result for each value of $N$ is printed to the terminal.

**Sample code:**

```fortran
implicit none
  integer, parameter :: dp = selected_real_kind(15, 307)
  integer :: i, N
  real(dp) :: a, b, h, x, integral

! Prompt for input
  write(*,'(A)', advance='NO') 'Enter integration limits a and b: '
  read(*,*)  a, b

  N_loop: do N = 2, 100, 2    ! Loop over the number of grid points
          h = (b - a) / N  ! Calculate the step size
      integral = 0.0_dp        ! Initialize the integral

! Compute the integral using the trapezoidal rule
    integral = 0.5_dp * (integrand(a) + integrand(b))
        do i = 1, N-1
          x = a + i * h
    integral = integral + integrand(x)
    end do
    integral = integral * h

! Output the result to screen
    write(*, 100) N, integral, a, b
100 format('N=',I3, 3X, 'Integral=', F12.10, 2X, 'a=', F6.3, 2X, 'b='&
        &, F6.3)
  end do N_loop

contains

! Define the integrand function
  real(dp) function integrand(x)
    real(dp), intent(in) :: x
    integrand = x**(5.0_dp/2.0_dp) * cosh(x)**2 * exp(-sqrt(x) *&
        & cos(x) - x**3 * sin(x**2)**2) * (1.0_dp + 0.5_dp * x +&
        & 0.2_dp * x**2 + 0.1_dp * x**4)**(-5)
  end function integrand
end program trapezoidal_integration
```

For $a = 0$ and $b = 2.0$, the terminal output has the form:

```
Enter the integration limits a and b: 0 2
N = 2 Integral=0.0366094915 a= 0.000 b= 2.000
N= 4 Integral=0.0403156797 a= 0.000 b= 2.000
 N= 6 Integral=0.0537433352 a= 0.000 b= 2.000
N= 8 Integral=0.0609615544 a= 0.000 b= 2.000
N= 10 Integral=0.0593255311 a= 0.000 b= 2.000
... ... ... ...
... ... ... ...
... ... ... ...
N= 20 Integral=0.0576783994 a= 0.000 b= 2.000
... ... ... ...
... ... ... ...
N=100 Integral=0.0576973546 a= 0.000 b= 2.000
```

## 6.10 Monte Carlo integration

*In this worksheet, we will numerically calculate the integral of a continuous function using the Monte Carlo (MC) method. We will compare the resulting value with those obtained using the trapezoidal and Simpson methods. Throughout, we will use double precision for numerical computation.*

The integral under study is defined as follows:

$$I \equiv \int_{a=0}^{b=1} f(x) \, dx, \tag{6.14}$$

where the integrand $f(x)$ is given by

$$f(x) = 5 \, e^{4x} (1 + \sinh(2x) \cosh(x))^{-2} - 5. \tag{6.15}$$

The MC method approximates the integral of a function $f(x)$ by the formula

$$\int_a^b f(x) \, dx \stackrel{\text{Monte Carlo}}{=} \frac{(b-a)}{M_{\text{MC}}} \sum_{k=1}^{M_{\text{MC}}} f(a + (b-a)r_k), \tag{6.16}$$

where $r_i$ is a randomly generated number between 0 and 1, and $M_{\text{MC}}$ is the chosen number of grid points. Generating a random number in a Fortran code is accomplished with CALL RANDOM_NUMBER(r), where the random number $r$ (returned by the subroutine) lies between 0 and 1.

**Tasks:**

Write a structured and well-commented Fortran program that computes the integral of equation (6.14) using the MC method, the Trapezoidal method, and the Simpson method. The program should include comments to explain each step of the calculation and the chosen methods. Finally, the numerical results obtained from each method will be compared with each other.

**Program design**

1. Use a statement function to define the integrand in equation (6.15).
2. The values for $a$, $b$, and $M_{MC}$ are all entered via keyboard input.
3. For the trapezoidal and Simpson integrations, use $N_{Trap} = 20$, and $N_{Simp} = 20$, respectively.
4. The MC integration can be carried out with the following Fortran syntax:

```
! Compute integral using Monte Carlo technique
average=0.0
do k=1,M_MC
call random_number(r) ! Generating random number r in
 ! the range [0,1]
x=a + (b-a)*r ! Converting r to be in the range [a,b]
average=average + f(x)! Summing all values of f(x)
end do
I_MC  =  (b-a)*average/FLOAT(M_MC)  ! Result  of  the
integral
```

5. Use the double precision format to utilize 64 bits to represent the floating-point numbers stored in `average`. The Fortran syntax for this is as follows:

```
integer, parameter :: dp = selected_real_kind(15, 307)
real (kind=dp) :: average
```

6. Assign names to all DO loops.
7. Run the program for $M_{MC}$ values of 20, 2000, and 2 000 000, and pay attention to the accuracy of the result obtained for the MC technique.

**Sample code:**

```
program MonteCarlo
  implicit none
! Declarations
  real :: a, b, x, r, h, I_MC, I_Trap, I_Simp, add, weight, average, f
  integer :: k, N_Trap = 20, N_Simp = 20, M_MC
! User input
  write(*,'(A)', advance='NO') 'Input integration limits a and b: '
  read(*,*) a, b
  write(*,'(A)', advance='NO') 'Input value for M_MC: '
  read(*,*) M_MC
! Compute integral using Monte Carlo technique
  average = 0.0
  do k = 1, M_MC
     call random_number(r)   ! Generating random number r in the range
                             ! [0,1]
     x = a + (b-a) * r       ! Converting r to be in the range [a,b]
     average = average + f(x) ! Summing all values of f(x)
  end do
  I_MC = (b-a) * average / REAL(M_MC)   ! Final result of the integral
  write(*,*) 'M_MC=', M_MC, 'a=', a, 'b=', b, 'I_MC=', I_MC
```

```
! Compute integral using the Trapezoidal rule
  h = (b-a) / REAL(N_Trap)
  add = 0.0
  do k = 1, N_Trap-1
     x = a + h * k
     add = add + f(x)
  end do
  I_Trap = h * (f(a) + 2.0*add + f(b)) / 2.0
  write(*,*) 'N_Trap=', N_Trap, 'a=', a, 'b=', b, 'I_Trap=', I_Trap
! Compute integral using Simpson's rule
  h = (b-a) / REAL(N_Simp)
  add = 0.0

  do k = 1, N_Simp-1
     x = a + h * k
     weight = 3 - (-1)**k
     add = add + weight * f(x)
  end do
  I_Simp = h * (f(a) + add + f(b)) / 3.0
  write(*,*) 'N_Simp=', N_Simp, 'a=', a, 'b=', b, 'I_Simp=', I_Simp
end program MonteCarlo

real function f(x)
    real, intent(in) :: x
    f = 5.0*EXP(4.0*x) / (1.0+SINH(2.0*x) * COSH(x))**2 - 5.0
end function f
```

The terminal dialog when running the code with M_MC=20 should appear as follows:

```
Input integration limits a and b: 0 1
Input value for M_MC: 20
M_MC= 20 a= 0.000 000 0 b= 1.000 000 0 I_MC= 1.160 369 2
N_Trap= 20 a= 0.000 000 0 b= 1.000 000 0 I_Trap= 1.361 134 8
N_Simp= 20 a= 0.000 000 0 b= 1.000 000 0 I_Simp= 1.362 211 3
```

where $I_{MC}$, $I_{Trap}$, and $I_{Simp}$ represent the results of the integral (6.14) obtained using the MC, trapezoidal, and Simpson methods, respectively. Note that the MC integration result varies randomly with each execution due to its probabilistic nature. Higher values of M_MC generally lead to more accurate results. For instance, with M_MC=2 000 000, the result of the MC integration is

```
M_MC= 2 000 000 a= 0.000 000 0 b= 1.000 000 0 I_MC= 1.361 519 2
```

This result closely aligns with those obtained using the Trapezoidal and Simpson methods.

## 6.11 Finding roots of a nonlinear equation

*Purpose: You will learn how to compute the roots of a nonlinear equation numerically using Newton's method.*

Given is the following nonlinear equation,

$$f(x) = e^x \ln(x) - x^2. \tag{6.17}$$

**Tasks:**

Write a Fortran program which uses Newton's method as described in section 4.7 to compute the root(s) of equation (6.17).

**Program design:**

1. Design the program such that $f(x)$ and $f'(x)$ are computed in different FUNCTION subprograms.
2. Stop Newton's iteration scheme if $|x_{i+1} - x_i| \leqslant \varepsilon$, where $\varepsilon = 10^{-6}$.
3. Start with an initial guess of $x = 0.8$ for finding the root.
4. Restrict the maximum number of iterations to 100.
5. Display the root value and the number of iterations on the screen.
6. Ensure that a message is printed to the standard output if the program fails to find the root(s) of the equation.
7. Generate a plot $f(x)$ and check whether or not your solution(s) for $f(x) = 0$ is (are) correct.

**Sample code:**

```
program solve_diff_eq
    implicit none

    real :: x, y, dx, y_analytic
    real, parameter :: x0 = 0.0, xfinal = 20.0, y0 = 6.75
    integer :: i, nsteps
    character(len=100) :: filename1, filename2

! Prompt user for step size
    write(*, "('Enter step size dx: ')", advance="no")
    read(*, *) dx

! Calculate number of steps
    nsteps = int((xfinal - x0) / dx)

! Open output files
    write(filename1, "(A, I0, A)") "output_eq1_dx", int(dx*1000.0),&
        & ".dat"
    open(unit=10, file=trim(filename1), status='replace')
```

```
    write(filename2, "(A, I0, A)") "output_eq2_dx", int(dx*1000.0),&
        & ".dat"
    open(unit=20, file=trim(filename2), status='replace')

! Initial condition
    x = x0; y = y0

! Compute numerical solutions
    do i = 1, nsteps
        y_analytic = -0.2 * cos(x) * cos(2.0 * x) - sin(x) + 7.0 * &
                     & cos(x)

! Write results to files
        write(10, *) x, y
        write(20, *) x, y_analytic

! Euler's method
        y = y + dx * (2.0 * cos(x)**3 * sin(x) - 1.0 - sin(x) * y) /&
            & cos(x)
        x = x + dx
    end do
    close(unit=10); close(unit=20)
end program solve_diff_eq
```

A sample output is as follows:

```
Root found: x=1.69 Number of iterations= 32
```

## 6.12 Ordinary differential equations

*Purpose: To illustrate how to solve an ordinary first-order differential equation numerically.*

Given is the ordinary first-order differential equation

$$\cos(x)\, y'(x) + \sin(x)y(x) = 2\cos^3(x)\sin(x) - 1, \tag{6.18}$$

where $0 \leqslant x \leqslant 20$. The initial condition is $y(0) = 6.75$. The analytic solution of equation (6.18) is given by

$$y_{\text{analytic}}(x) = -0.2\cos(x)\cos(2x) - \sin(x) + 7\cos(x). \tag{6.19}$$

**Tasks:**

Write a structured Fortran program to numerically solve equation (6.18) using Euler's method as described in section (4.8.1). Compare the numerical results graphically with the analytic solution obtained from equation (6.19).

**Program design:**

1. Use a logical 'IF' statement to assign a numerical value to $x \in [0, 20]$. Prompt the user to input the desired step size $\Delta x$.

2. Employ the 'OPEN' statement to write the results of equations (6.18) and (6.19) for $0 \leqslant x \leqslant 20$ to external output files. Advanced used may use the following Fortran syntax:

```
! Open output files
write(filename1, "(A, I0, A)") "output_eq1_dx", int
(dx*1000.0),&
 & ".dat"
open(unit=10, file=trim(filename1), status='replace')
write(filename2, "(A, I0, A)") "output_eq2_dx", int
(dx*1000.0),&
 & ".dat"
open(unit=20, file=trim(filename2), status='replace')
```

The first `write` statement constructs a filename 'filename1' by concatenating the prefix `output_eq1_dx` with an integer representation of $dx$ scaled by 1000, followed by the extension `.dat`.

3. Generate a plot depicting $y(x)$.

4. Include on the same plot $y_{analytic}(x)$ for $\Delta = 0.01$ and $\Delta = 0.001$.

**Sample code:**

```
program projectile
 implicit none
 real :: m, rho, Re, rad, deltaT, z_i, z_ip1=0, v_i, v_ip1, b, C_d
 real :: time=0.0, g, b_v, area, dragF
 real, parameter :: pi=ACOS(-1.0)
 data g/9.81/, rad/5.0E-03/, rho/1.5E03/, Re/1.5E04/, m/5.0E-02/

 open(unit=10, file='position.dat', status='unknown')
 open(unit=20, file='speed.dat',    status='unknown')

 write(*, 50, advance='NO')
50  format(1x, "Object's initial position (in meter): ")
    read (*, *) z_i
    write(*, 51, advance='NO')
51  format(1x, "Object's initial speed (in meter/seconds): ")
    read (*, *) v_i
    write(*, 52, advance='NO')
```

```
52  format(1x,"Stepsize delta t (in seconds): ")
    read (*, *) deltaT

! Object's initial position and speed
    write(*,54) time, z_i;  write(*,55) time, v_i
54  format(1x, 't=', F4.2, ' seconds,', 4x, 'z(t)=', F7.2, ' meter')
55  format(1x, 't=', F4.2, ' seconds,', 4x, 'v(t)=', F7.2, ' meter &
        &/seconds')

 C_d  = compute_drag(Re)               ! Compute drag coefficient
 area = compute_cross_section(rad, pi) ! Compute cross section of
                                       ! particle (in m^2)
 b    = compute_constant_b(rho, C_d, area) ! Compute b (in kg/m)
```

This code automatically generates the output files 'output_eq1_dx10.dat' and 'output_eq2_dx10.dat' and writes the results into these files, where '10' is determined by 'int(dx*1000.0)'. The analytic and numerical solutions of Equations (6.18) and (6.18) are shown graphically in figure 6.4.

## 6.13 Projectile in a viscous medium

*Purpose: To Illustrate how to solve ordinary differential equations numerically.*

A spherical projectile moves vertically in a viscous medium near the surface of the Earth. It experiences two forces, the gravitational attraction of the Earth $(-m\,g)$ and the viscous force $F_v$ from the medium in which it moves. The former is directed downward (negative $z$) direction and the latter is directed opposite to the velocity $v$.

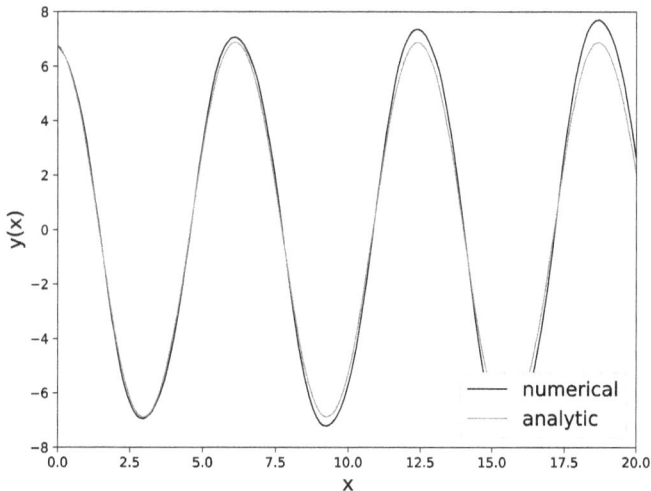

**Figure 6.4.** Numerical and analytic solutions of equations (6.18) and (6.19) obtained using the illustrated Fortran code with a step size of $\Delta = 0.01$. A step size of $\Delta = 0.001$ would result in perfect agreement with the analytic curve.

Usually, the magnitude of the viscous force is a function of the speed with which the projectile moves, symbolically, $F_v = f(v)$. Assuming that $f = b\,v^2$ ($b > 0$ denotes a positive constant), the projectile's equation of motion is then given by

$$m\frac{d^2z}{dt^2} = -m\,g - b\,\frac{dz}{dt}\left|\frac{dz}{dt}\right|. \tag{6.20}$$

The parameters in this equation are the projectile's mass, $m = 50$ g, and the gravitational acceleration, $g = 9.81$ m s$^{-2}$. The constant $b$ is given by

$$b = \frac{1}{2}\,\rho\,C_d\,A, \tag{6.21}$$

where $\rho$ is the density of the viscous medium, $C_d$ is the drag coefficient,

$$C_d \approx \frac{24}{Re} + \frac{6}{1 + \sqrt{Re}} + 0.4, \tag{6.22}$$

$A$ is the cross-sectional area, and $Re$ denotes the Reynolds number[2].

**Tasks:**

Write a structured Fortran program which solves the nonlinear second-order differential equation (6.20) using the midpoint method as described in section 4.8.2 for given values of $m$, $g$, $\rho = 1500$ kg m$^{-3}$, and $Re = 1.5 \times 10^4$. The radius of the projectile is $r = 5$ mm. The initial conditions of the projectile are $z_0 \equiv z(t = 0) = 0$ m and $v_0 \equiv \dot{z}(t = 0) = 5$ m s$^{-1}$.

**Program design:**

1. The formatted screen output should be as follows:

```
Projectile's initial position (in meter):
Projectile's initial speed (in meter/seconds):
Stepsize delta t (in seconds): 0.0001
t=0.00 seconds, z(t)= 0.00 meter
t=0.00 seconds, v(t)= 5.00 meter/seconds
```

2. Assume a temporal step size of $\Delta t = 0.001$ s.
3. Use FUNCTION subprograms to compute $C_d$, the area (cross section) of the spherical projectile, and the constant $b$.
4. Generate a plot which shows the projectile's position $z(t)$ and speed $v(t)$ as a function of time, $t$.
5. Explore the dependence of your numerical solutions on the value chosen for $\Delta t$ by running your code for time steps $\Delta t = 0.1$ s, 0.01 s, and 0.001 s. Show the results in the plots that you just generated.
6. Additional tasks: Modify your program to include the Runge–Kutta method as described in section 4.8.3 via the expressions in equation (4.119). Show the

---

[2] The Reynolds number is a criterion of whether fluid, liquid, or gas flow is steady (laminar) or unsteady (turbulent). If the Reynolds number is less than around 2000, flow is generally laminar. For Reynolds numbers greater than around 2000, flow is usually turbulent.

results for $z(t)$ and $v(t)$ graphically for the midpoint method and the Runge–Kutta method in one plot and comment on any differences.

**Sample code:**

```
z_loop: do while (z_ip1 >= 0.0)          ! Solve differential equation
        z_ip1 = z_i + v_i * deltaT
        drag_force: if (ABS(v_i) < 3.0) then
                        b_v = b
                    else
                        b_v = (1.0 + TANH((ABS(v_i)-3.0))) * b
                    end if  drag_force
        dragF = - b_v * v_i * ABS(v_i)
        v_ip1 = v_i - deltaT * (m*g - dragF) / m
        time  = time + deltaT
        write(10, 53) time, z_ip1  ! Output z(t)
        write(20, 53) time, v_ip1  ! Output v(t)
   53   format(1x, F9.5, 4x, F10.5)
        z_i = z_ip1;  v_i = v_ip1  ! Update position and speed
    end do z_loop

    close(unit=10);  close(unit=20)

contains

 real function compute_drag(Re)
     implicit none
     real, intent(in) :: Re
     compute_drag = 24.0 / Re + 6.0 / (1.0 + SQRT(Re)) + 0.4
 end function compute_drag

 real function compute_cross_section(radius, pi)
     implicit none
     real, intent(in) :: radius, pi
     compute_cross_section = radius**2 * pi
 end function compute_cross_section

 real function compute_constant_b(rho, C_d, area)
     implicit none
     real, intent(in) :: rho, C_d, area
     compute_constant_b = rho * C_d * area / 2.0
 end function compute_constant_b
end program projectile
```

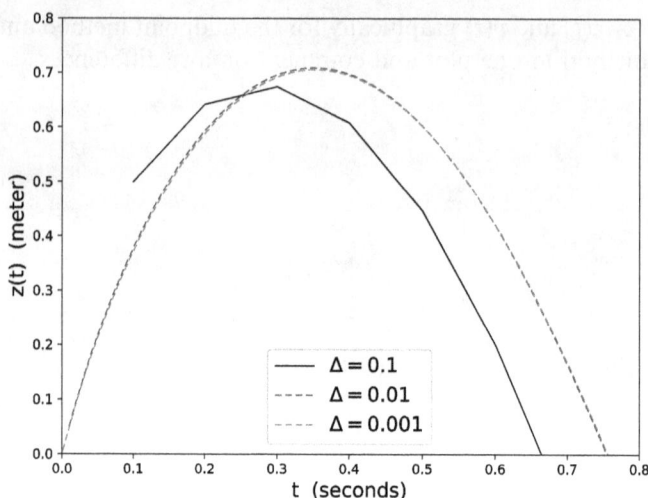

**Figure 6.5.** Path of a projectile moving in a viscous medium, computed from equation (6.20) for temporal step sizes $\Delta = 0.1$ s, 0.01 s, and 0.001 s. Numerically acceptable solutions are obtained for $\Delta = 0.01$ s and 0.001 s.

Numerical solutions generate by this code for different temporal step sizes are shown in figure 6.5.

## 6.14 Damped harmonic oscillator

*Purpose: To illustrate the methods and techniques used to numerically solve higher-order differential equations. This will include a step-by-step approach to setting up the equations, choosing appropriate numerical methods, and implementing the solution in a computational program.*

Given is an object of mass $m$ attached to a spring with spring constant $\kappa$. The object oscillates back and forth in the $x$-direction. The motion of $m$ is damped by a frictional, velocity-dependent force $-\beta \dot{x}$, where $\beta$ is a constant. The equation of motion of $m$ is thus given by

$$m\,\ddot{x} + \kappa\,x + \beta\,\dot{x} = 0, \tag{6.23}$$

with $m = 2 \times 10^4$ g and $\beta = 14$ kg s$^{-1}$.

**Tasks:**

Write a structured and well-commented Fortran program that solves equation (6.23) numerically via one or more of the methods described in sections 4.8.1–4.8.3. You will do this for both $\beta = 0$ (i.e., no damping) and $\beta \neq 0$, and illustrate the results graphically.

From your results, determine an approximate value for the amplitude, $A$, and the period of oscillation, $T$.

The initial conditions are $x(0) = 4$ m and $\dot{x}(0) = -150$ cm s$^{-1}$. The acceleration of $m$ at $t = 0$ is $-10^3$ cm s$^{-12}$.

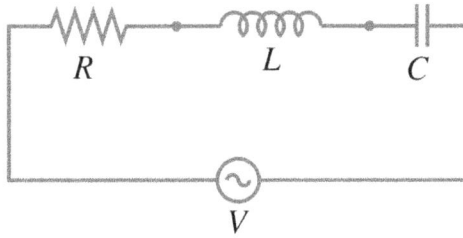

**Figure 6.6.** Illustration of a RLC circuit.

**Program design:**
1. Design your code such that the user is prompted to input a value for the time step $\Delta t$. The numerical results for $x(t)$ and $v(t)$ are to be written to output files.
2. Generate a plot for $x(t)$ and $v(t)$ for both the undamped as well as the damped case for $0 \leqslant t \leqslant 15$ s.
3. From your program, also numerically determine the force, $F_3$, that acts on $m$ at $t = 0.3$ s (i.e., determine $F_3(0.3)$).

## 6.15 RLC circuit

*Purpose: Illustrate how to solve higher-order differential equations numerically and learn the 90 module feature.*

An RLC circuit [2, 3] is an oscillating circuit consisting of a resistor (R), capacitor (C), and inductor (L) connected in series, as shown in figure 6.6. The capacitor is charged initially. The voltage of this charged capacitor causes a current ($I = dq/dt$) to flow in the inductor to discharge the capacitor. Once the capacitor is discharged, the inductor resists any change in the current flow, causing the capacitor to be charged again with the opposite polarity. The voltage in the capacitor eventually causes the current flow to stop and then flow in the opposite direction. The result is an oscillating electric current. The differential equations which describes the flow of the electric current throughout the RLC circuit is given by

$$\frac{d^2q}{dt^2} = -\frac{R}{L}\frac{dq}{dt} - \frac{q}{L\,C}, \tag{6.24}$$

where $L = 0.012$ H and $C = 1.0 \times 10^{-5}$ F[3].

**Tasks:**
Write equation (6.24) as a system of coupled first-order differential equations. For this system, write a structured Fortran program that numerically solves this system of equations using one or more of the methods described in sections 4.8.1 to 4.8.3, for times $0 \leqslant t \leqslant t_{\text{final}}$, where $t_{\text{final}} = 5 \times 10^{-3}$ s. The initial conditions for the electric charge and electric current are $q(t = 0) = 1.6 \times 10^{-5}$ Coulomb and $\frac{dq}{dt}\big|_{t=0} = I(t = 0) = 0.01$ A, respectively.

---

[3] H=$\Omega$ s, F=s/$\Omega$.

**Program design:**
1. Choose $\Delta t = t_{\text{final}}/500$ for the temporal step size.
2. Use a FUNCTION to evaluate the right-hand side of equation (6.24).
3. Utilize the Fortran module feature to assign values to $L$, $C$, $I(0)$, and $q(0)$.
4. Design your code so that the user is prompted to input the numerical value for $R$ from the keyboard.
5. Generate a plot showing $q$ as a function of $t$ for $R = 1.5$ $\Omega$, $R = 5$ $\Omega$, and $R = 50$ $\Omega$ (all on a single plot).

**Sample code:**

```fortran
module InputData
   implicit none
   real :: L = 0.012      ! Inductance (H)
   real :: C = 1.0E-05    ! Capacitance (F)
   real :: I_1 = 0.01     ! Initial current (A)
   real :: q_1 = 1.6E-05  ! Initial charge (Coulomb)
end module InputData

program RLC_circuit
   use InputData
   implicit none
   real    :: R, q_2, I_2, t, t_final, dt
   integer :: n_step

! Parameters initialization
   t_final = 5.0E-03    ! Final time (s)
   n_step = 500         ! Number of time steps
   dt = t_final / real(n_step) ! Time step size
! File handling
   open(unit=22, file='RLC.dat', status='unknown')

! Prompt user for input value of R
   write(*,'(A)', advance='NO') 'Input value for R (in Ohm): '
   read(*, *) R

! Display simulation parameters
   print*, 'Simulation parameters:'
   print*, 't_final =', t_final, 's'
   print*, 'n_step =', n_step
   print*, 'dt =', dt
   print*, 'R, L, C =', R, L, C

! Initialize time
   t = 0.0

! Time integration loop
time:   do while (t <= t_final)
          t = t + dt
          q_2 = q_1 + I_1 * dt
          I_2 = I_1 - (q_1 / (L * C) + (R / L) * I_1) * dt
```

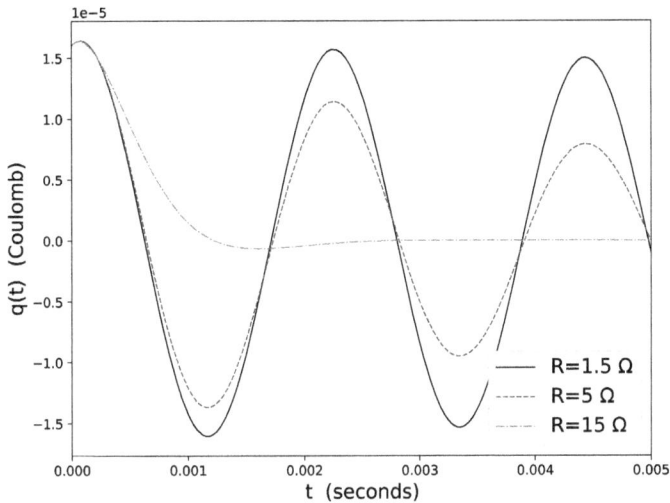

**Figure 6.7.** Temporal behavior of electric charge in an RLC circuit, computed from equation (6.24) for $R = 1.5\ \Omega$, $5\ \Omega$, and $50\ \Omega$.

```
! Write output to file
        write(22, *) t, q_2

! Update variables for next time step
        q_1 = q_2; I_1 = I_2
    end do time

    close(22)
end program RLC_circuit
```

Numerical solutions generated by this code for different values of $R$ are shown in figure 6.7.

## References

[1] Etter D M 1995 *Fortran 90 for Engineers* 1st edn (New York: Wiley)

[2] Edminister J and Nahvi M 2013 *Schaum's Outline of Electromagnetics* 4th edn (Schaum's Outline Series) (New York: McGraw-Hill Education)

[3] Edminister J and Nahvi M 2011 *Schaum's Outline of Electric Circuits* 5th edn (Schaum's Outline Series) (New York: McGraw-Hill Education)

**IOP** Publishing

# Introduction to Computational Physics for Undergraduates
## (Second Edition)

**Omair Zubairi and Fridolin Weber**

# Chapter 7

## Homework assignments

## 7.1 Fresnel coefficients

To study the reflection and transmission of light at a material interface (e.g. air-glass), one typically examines three distinct waves traveling in the directions depicted in the figure 7.1, where $E^{(p)}$ and $E^{(s)}$ denote the components of the electric field in the parallel ($p$) and vertical ($s$) directions. The subscripts $r$, $i$, and $t$ stand for reflected, incident, and transmitted, respectively. The index of refraction $n_i$ charac-terizes the material on the left-hand side and $n_t$ characterizes the material on the right-hand side of the vertical axis. The incident plane wave makes an angle $\theta_i$ with the normal to the interface, the reflected wave makes an angle $\theta_r$ with the interface normal, and the transmitted plane wave makes an angle $\theta_t$ with the interface normal. The ratio of the reflected and transmitted electric field components to the incident field components are specified by the following coefficients, called Fresnel coefficients:

$$r_s \equiv \frac{E_r^{(s)}}{E_i^{(s)}} = \frac{n_i \cos\theta_i - n_t \cos\theta_t}{n_i \cos\theta_i + n_t \cos\theta_t}, \tag{7.1}$$

$$t_s \equiv \frac{E_t^{(s)}}{E_i^{(s)}} = \frac{2n_i \cos\theta_i}{n_i \cos\theta_i + n_t \cos\theta_t}, \tag{7.2}$$

$$r_p \equiv \frac{E_r^{(p)}}{E_i^{(p)}} = \frac{n_i \cos\theta_t - n_t \cos\theta_i}{n_i \cos\theta_t + n_t \cos\theta_i}, \tag{7.3}$$

$$t_p \equiv \frac{E_t^{(p)}}{E_i^{(p)}} = \frac{2n_i \cos\theta_i}{n_i \cos\theta_t + n_t \cos\theta_i}. \tag{7.4}$$

doi:10.1088/978-0-7503-6493-5ch7

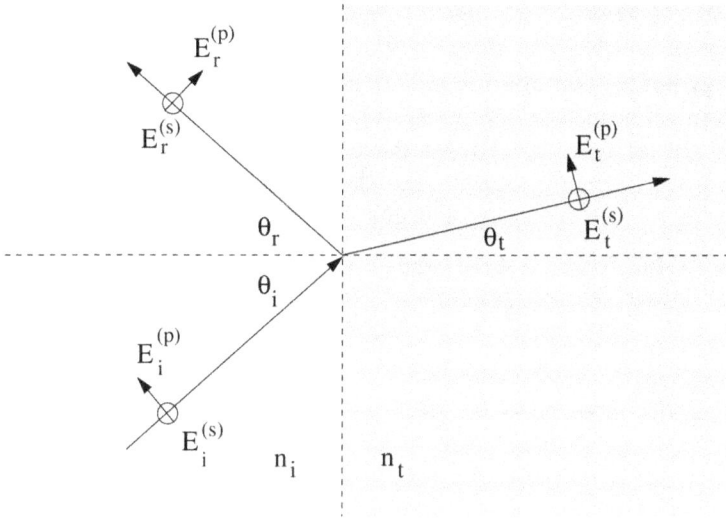

**Figure 7.1.** Incident (*i*), reflected (*r*), and transmitted (*t*) plane waves at a material (e.g., air-glass) interface.

The Fresnel coefficients allow one to easily connect the electric field amplitudes on the two sides at the material interface. They also keep track of phase shifts at a boundary.

**Tasks:**

Write a structured Fortran program which computes (and outputs) the Fresnel coefficients given in equations (7.1) to (7.4) for $0 \leqslant \theta_i \leqslant 90°$.

**Program design:**

1. Use $n_i = 1$ and $n_t = 1.5$ (air-glass interface), and a fixed angle of $\theta_t = 5°$ for the transmitted light ray.
2. The program should also compute (and output) the reflectances $R_s \equiv r_s^2$ and $R_p \equiv r_p^2$, and the transmittances $T_s \equiv 1 - R_s$ and $T_p \equiv 1 - R_p$, for $0 \leqslant \theta_i \leqslant 90°$.
3. Generate a plot which shows the Fresnel coefficients $r_s$, $t_s$, $r_p$, and $t_p$ for $0 \leqslant \theta_i \leqslant 90°$.
4. Generate a plot which shows $R_s$, $R_p$, $T_s$ and $T_p$ for $0 \leqslant \theta_i \leqslant 90°$.

## 7.2 Semiempirical mass formula of atomic nuclei

The Bethe–Weizsäcker mass formula, also known as the semiempirical mass formula, expresses the mass of an atomic nucleus, $M$, as a function of the proton number $Z$ and the nucleon (baryon) number $A$ as follows:

$$M(Z, A) = ZM_H + (A - Z)M_n - a_V A + a_S A^{2/3} + a_C Z^2 A^{-1/3} \\ + a_A (Z - A/2)^2 A^{-1}. \tag{7.5}$$

The quantities $a_V = 15.8$ MeV, $a_S = 18.3$ MeV, $a_C = 0.71$ MeV, and $a_A = 92.7$ MeV are experimentally determined constants. (Use $M_n = 939.57$ MeV and $M_H = 938.78$ MeV for the mass of a neutron and the mass of a hydrogen atom, respectively.)

The energy per baryon of an atomic nucleus, typically referred to as the nuclear binding energy per nucleon, is the average energy required to remove a nucleon (either a proton or a neutron) from the nucleus. This quantity is key in nuclear physics because it provides insight into the stability of a nucleus and the energy released during nuclear reactions. The energy per baryon, $E_B(Z, A)$, is derived from equation (7.5) as follows:

$$E_B(Z, A) \equiv (M(Z, A) - ZM_H - (A - Z)M_n) A^{-1} \qquad (7.6)$$

$$= -a_V + a_S A^{-1/3} + a_C Z^2 A^{-4/3} + a_A (Z - A/2)^2 A^{-2} \qquad (7.7)$$

$$\equiv E_{volume} + E_{surface} + E_{coulomb} + E_{asymmetry}, \qquad (7.8)$$

where

$$E_{volume} = -a_V, \qquad (7.9)$$

$$E_{surface} = a_S A^{-1/3}, \qquad (7.10)$$

$$E_{coulomb} = a_C Z^2 A^{-4/3}, \qquad (7.11)$$

$$E_{asymmetry} = a_A (Z - A/2)^2 A^{-2}, \qquad (7.12)$$

denote the volume energy, surface energy, Coulomb energy, and asymmetry energy of a given atomic nucleus, respectively.

**Tasks:** Write a structured Fortran program to compute the properties of an atomic nucleus, including $E_{Volume}$, $E_{Surface}$, $E_{Coulomb}$, $E_{Asymmetry}$, $M(Z, A)$, and $E_B(Z, A)$, based on user-provided proton number ($Z$) and baryon number ($A$).

**Program design:**

1. Ensure the program contains a preamble.
2. Comment the program code effectively.
3. Define $Z$, $A$, and $N$ as integers.
4. Accept keyboard input for the proton number ($Z$) and the baryon number ($A$), using the advance='NO' option.
5. Utilize a DO loop to prompt the user for $Z$ and $A$ values for seven different atomic nuclei sequentially. The program should then calculate the properties of each nucleus, display the results, and prompt the user for the next nucleus's $Z$ and $A$ values. This process repeats until the seventh nucleus's input is provided.
6. The screen output for a nucleus such as $^{244}_{94}$Pu should resemble the following format:

```
Input proton number (Z): 94
Input baryon number (A): 244
```

```
N = 150 (Number of neutrons)
Z = 94 (Number of protons)
A = 244 (Baryon number)
M(Z,A)= 227 342.016 MeV
E_B= -7.53611803    MeV
E_volume = -15.8000002    MeV
E_surface = 2.928 561 93  MeV
E_coulomb = 4.114 598 75  MeV
E_asymmetry= 1.220 720 17 MeV
```

Use this program to determine the masses and binding energies of the following atomic nuclei in a single program run[1]:

$$^{70}_{32}\text{Ge}, \quad ^{179}_{72}\text{Hf}, \quad ^{38}_{18}\text{Ar}, \quad ^{56}_{26}\text{Fe}, \quad ^{40}_{20}\text{Ca}, \quad ^{235}_{92}\text{U}, \quad ^{238}_{92}\text{U}.$$

## 7.3 Magnetic permeability

The magnetic field around a wire carrying an electric current $I$ is given by

$$B(r) = \mu_0 I/(2\pi r),$$

where $r$ is the distance measured from the wire and $\mu_0$ is the magnetic permeability.

Experimental results of the measured magnetic field, $B$, as a function of $r$ is given in table 7.1. The current is kept constant at 2.7 Amperes. Create a file named Br. dat that contains the data shown in table 7.1. This file should consist of two columns: the first column contains the $r$ values and the second column contains the $B(r)$ values.

**Tasks:**

Write a structured Fortran program which determines the value of $\mu_0$ using the linear (i.e. $f(x) = a + bx$) least-squares method as described in section 4.1.1. (The units of the permeability are $\mu$T m/A.)

**Table 7.1.** Experimental measurement of magnetic field $B$ at a certain distance from a current carrying wire.

| Data Points | 1 | 2 | 3 | 4 | 5 | 6 | 7 | 8 | 9 | 10 |
|---|---|---|---|---|---|---|---|---|---|---|
| $r$ (cm) | 10.0 | 20.0 | 30.0 | 40.0 | 50.0 | 60.0 | 70.0 | 80.0 | 90.0 | 100.0 |
| $B(\mu\text{T})$ | 5.4 | 2.7 | 1.8 | 1.4 | 1.0 | 0.88 | 0.76 | 0.63 | 0.59 | 0.51 |

---

[1] We use the notation $^A_Z\text{X}$ for an atomic nucleus of type X, where $A = N + Z$ is the number of nucleons (baryon number), $N$ the number of neutrons, and $Z$ the number of protons of nucleus X.

**Program design:**
1. The code should include a preamble.
2. Experimental data will be read from the file `Br.dat` using a SUBROUTINE named `INPUTBr`.
3. Compute the coefficients $a$ and $b$ within a SUBROUTINE called `SUMMATIONS`.
4. Compute the permeability using a FUNCTION named `PERMEABILITY`. The units of permeability are in $\mu$T m/A.
5. Display the permeability result on standard output in the main program.
6. Generate a plot that displays the experimental data from `Br.dat` along with the best-fit line obtained using the least-squares linear model.

## 7.4 Fourier sine transforms

The Fourier sine transform of a function $f(t)$ is a mathematical operation used in signal processing and mathematical analysis to decompose a function into its constituent sinusoidal components. Unlike the more familiar Fourier transform, which uses complex exponentials, the Fourier sine transform exclusively employs sine functions. Here, we consider a Fourier sine transform given by

$$F(\omega) = \int_0^\infty dt\, f(t)\, \sin(\omega t), \qquad (7.13)$$

where $t$ is the time in seconds and $\omega$ denotes the angular frequency in s$^{-1}$. The function $f(t)$ is chosen as

$$f(t) = e^{-b\,t}, \qquad (7.14)$$

where $b = 2$ s$^{-1}$.

**Tasks:**
Write a structured Fortran program that computes $F(\omega)$ for $\omega$ values $0 \leqslant \omega \leqslant 10$ s$^{-1}$. The numerical results are to be compared with the exact (analytic) result of the Fourier sine transform of $f(t)$ given by

$$F_b(\omega) = \frac{\omega}{\omega^2 + b^2}. \qquad (7.15)$$

**Program design:**
1. Include a short preamble at the beginning of your program.
2. There is no keyboard input.
3. Use FUNCTION subprograms to define the integrand of the integral of equation (7.13) and $F_b(\omega)$ of equation (7.15).
4. Use the Trapezoidal rule to compute the integral of equation (7.13). Use $t(\infty) = 6/b$ and $N = 100$ for the number of grid points on the $t$-axis.
5. For $\omega$, use a DO loop to move from 0 to 10 s$^{-1}$ in steps of $\Delta\omega = 10$ s$^{-1}/M$, where $M = 300$.
6. The numerical results for $F(\omega)$ are to be written to an external data file named `FourierTrNum.dat`.

7. The analytic results are to be written to `FourierTrExact.dat`.
8. Show your numerical and analytic results graphically in the same plot.

## 7.5 Kinetic friction

An object with a mass of 5.5 kg slides from rest down an inclined plane. The plane makes an angle of $\theta = 30°$ with the horizontal and is $s = 72$ m long. The speed of the object at the bottom of the plane is $v = 16.7$ m s$^{-1}$ and follows from

$$v = \sqrt{2g(\sin\theta - \mu\cos\theta)s}\,,$$

where $g = 9.81$ m s$^{-12}$ and $\mu$ is the coefficient of kinetic friction between the plane and the object.

**Tasks:**

Write a structured and well-commented Fortran program which uses Newton's numerical root-finding method as described in section 4.7 to determine $\mu$.

**Program design:**

1. The value of $v$ is keyboard input and the maximum number of iterations is 20.
2. Use an initial value of $\mu = 0.5$ to start the root-finding algorithm.
3. Terminate the calculations if $\Delta \equiv ||\mu_{i+1}|-|\mu_i|| < 10^{-5}$.
4. Use FUNCTIONs to determine $F(\mu)$ and $dF(\mu)/d\mu$.
5. For each iteration step, write $\Delta \& \mu$ to standard output.

## 7.6 Compton scattering

Suppose that x-rays of $E = 100$ keV energy are incident on a target and undergo so-called Compton (i.e. electron–photon) scattering. The scattering process can be described by the following formula,

$$\cos\phi = \frac{E^2 - E'^2 + K^2(1 + 2E_0/K)}{2EK\sqrt{1 + 2E_0/K}},$$

where $\phi = 73°$ is the angle of the recoiling electrons, $E'$ is the energy of the scattered x-rays, $K = 2.5$ keV, and $E_0 = 511$ keV is the restmass of an electron.

**Tasks:**

Write a structured Fortran program which uses the numerical root-finding method as described in section 4.7 to determine $E'$.

**Program design:**

1. The value of $E$ is keyboard input.
2. Limit the maximum number of iterations to 20.
3. Use an initial value of $E' = 10$ keV to start the root-finding algorithm.
4. Terminate the calculations if $\Delta \equiv ||E'_{i+1}|-|E'_i|| < 10^{-5}$.
5. Use FUNCTIONs to determine $F(E')$ and $dF(E')/dE'$.
6. For each iteration step, write $\Delta$ and $E'$ to standard output.

## 7.7 Radioactive decay

Given are three radioactive atomic nuclei, $A$, $B$, and $C$, which decay according to the following radioactive decay chain:

$$A \longrightarrow B \longrightarrow C.$$

The decay is described by the following system of coupled differential equations,

$$dA/dt = -k_A A, \quad dB/dt = k_A A - k_B B, \quad dC/dt = k_B B, \qquad (7.16)$$

where $k_A$ and $k_B$ are decay constants, and $A(t)$, $B(t)$, and $C(t)$ are the number of nuclei of each species present. The differential equations are coupled because each of the second and third of the them involves two of the dependent variables.

The initial values are $A(0) = A_0$, $B(0) = B_0$, and $C(0) = C_0$. The differential equations then have a unique solution, and that solution will depend on the parameters $k_A$ and $k_B$ and on the three initial values.

**Tasks:**

Write a structured and well-commented Fortran program which solves the radioactive decay equation (7.16) using Euler's method as described in section 4.8.1.

**Program design:**

1. Compute solutions for $A(0) = 1000$, $B_0 = C_0 = 0$, $k_A = k_B = 0.1$ and a time step of $\Delta t = 0.25$ s.
2. Generate a plot which shows $A(t)$, $B(t)$, $C(t)$ for $0 \leqslant t \leqslant 50$ s.
3. Comment on the results for $A + B + C$. How are they connected with equation (7.16)?

## 7.8 Halley's comet

Halley's comet last reached perihelion (its point of closest approach to the Sun at the origin) on February 9, 1986. Its position and velocity components at this time were:

$$\vec{r}(0) \equiv (x(0), y(0), z(0)) = (0.325\ 514, -0.459\ 460, 0.166\ 229),$$
$$\vec{v}(0) \equiv (v_x(0), v_y(0), v_z(0)) = (-9.096\ 111, -6.916\ 686, -1.305\ 721),$$

respectively, with position in AU (Astronomical Units, the unit of distance being equal to the major semiaxis of the Earth's orbit about the Sun). The time is measured in years. In this unit system, its three-dimensional equations of motion are as follows:

$$\frac{d^2x}{dt^2} = -\frac{\mu x}{r^3}, \qquad \frac{d^2y}{dt^2} = -\frac{\mu y}{r^3}, \qquad \frac{d^2z}{dt^2} = -\frac{\mu z}{r^3}, \qquad (7.17)$$

where $\mu = 4\pi^2$, and $r = \sqrt{x^2 + y^2 + z^2}$.

**Tasks:**

Write a structured Fortran program which solves equation (7.17) numerically via one or more of the methods described in sections 4.8.1 to 4.8.3 to find and illustrate the results graphically via a plot of the $yz$ projection (i.e. $z$ versus $y$) of the orbit of Halley's comet.

Use your numerical solution to determine Halley's maximum distance (at aphelion), the comet's period of revolution, and the time needed to return to perihelion. (Hint: plot $r(t)$.)

Using your results, determine the best estimate of the calendar date of the comet's next perihelion passage?

**Program design:**
1. Use the `Parameter` statement for $\pi$.
2. Use a time step of $\Delta t = 0.001$ and have user input for the final time.
3. Have your program also generate plots for both the $xy$ (i.e. $y$ versus $x$) and $xz$ (i.e. $z$ versus $x$) projections.

**Investigate your own comet**

Lucky you! The night before your birthday in 1997 you set up your telescope on a nearby mountaintop. It was a clear night, and at 12:30 am you spotted a new comet. After repeating the observation on successive nights, you were able to calculate its Solar System coordinates $r_0 = (x(0), y(0), z(0))$ and its velocity vector $v_0 = (v_x(0), v_y(0), v_z(0))$ on that first night. Using this information, determine this comet's:

  – Perihelion (point nearest Sun) and aphelion (farthest from Sun),
  – Its velocity at perihelion and at aphelion,
  – Its period of revolution about the Sun, and
  – Its next two dates of perihelion passage.

Using length-time units of AU and earth years, the comet's equations of motion are given in equation (7.17) with $\mu = 4\pi^2$. For your personal comet, start with random initial position and velocity vectors with the same order of magnitude as those of Halley's comet. Repeat the random selection of initial position and velocity vectors, if necessary, until you get a nice-looking eccentric orbit that goes well outside the Earth's orbit (like real comets do).

## 7.9 Rocket equation

The equation that describes the motion of a rocket is given by the following differential equation:

$$\frac{dv}{dt} = \frac{R\,u_{ex}}{m_i - Rt} - g. \tag{7.18}$$

This equation is called the rocket equation. The quantity $F_{th} = Ru_{ex}$ is the force exerted on the rocket by the exhausting fuel, and is called the thrust ($R$ denotes the burn rate, $u_{ex}$ is the speed at which the fuel is exhausted relative to the rocket). The payload of a rocket is defined as $m_f/m_i$, where $m_f$ denotes the rocket's final mass after all the fuel has been burned, and $m_i$ is the rocket's initial mass. The quantity $g$ denotes the gravitational acceleration ($9.81$ m s$^{-2}$). The Saturn-V rocket used in the Apollo moon-landing program had an initial mass of $m_i = 2.85 \times 10^6$ kg, a payload of 27%, a burn rate of $1.384 \times 10^4$ kg s$^{-1}$, and a thrust of $3.4 \times 10^7$ N.

**Tasks:**

Using these values, write a Fortran program that solves the rocket equation (7.18) numerically via one or more of the methods described in sections 4.8.1 to 4.8.3.

**Program design:**

1. Use the PARAMETER statement to assign a value to the gravitational acceleration in your program.
2. Use the DATA statement to assign values to the rocket data.
3. The temporal step size, $dt$, is to be read from the keyboard.
4. Compute the rocket's final mass, $m_f$, in a FUNCTION subprogram called RMASS_F.
5. Compute the rocket's burn time, given by $t_b = (m_i - m_f)/R$, in a FUNCTION subprogram called BURNT.
6. Print the values of $u_{ex}$, $m_f$, $t_b$ on the terminal screen. The output should look as follows:

| $u_{ex}$ (km s$^{-1}$) | $m_f$ (tons) | $t_b$ (s) |
| --- | --- | --- |
| ... | ... | ... |
| ... | ... | ... |

7. Solve the rocket equation in a SUBROUTINE called VELOCITY and write the results to an external data file.
8. In the same subroutine, compute the exact value of the rocket's velocity at time $t$, which is given by

$$v(t) = -u_{ex}\ln(1 - Rt/m_i) - gt \qquad (7.19)$$

and save the results to another external data file.
9. Generate a plot that compares the numerical solution of the rocket equation (computed for time steps of 0.1, 1, 10, and 20 seconds) with the exact (i.e. analytical) solution given by equation (7.19).

## 7.10 Hydrostatic equilibrium and relativistic stars

Galaxies are filled with billions of so-called compact stars. Such objects are as massive as our Sun but have radii that are just around 10 kilometers[2]. The densities inside compact stars are therefore 10–20 times higher than the density of atomic nuclei! In this assignment we will compute the mass–radius relationship of such stars, which follows from the first-order differential equation

---

[2] The mass of our Sun is $M_\odot = 2 \times 10^{30}$ kg, its radius is around $R \sim 700\,000$ km.

$$\frac{dP(r)}{dr} = -\frac{[\varepsilon + P(r)][4\pi r^3 P(r) + m(r)]}{r^2\left(1 - \dfrac{2m(r)}{r}\right)}, \tag{7.20}$$

where $P$, $\varepsilon$, and $m$ denote pressure, energy density, and mass of the mass distribution at a radial distance $r$ from the center of the star, where we have used the geometrical units of $G = c = 1$. Equation (7.20) follows from Albert Einstein's theory of general relativity and is known as the Tolman–Oppenheimer–Volkoff (TOV) equation, as first derived in [3, 4]. The Newtonian limit of (7.20) is given by ($P \ll \varepsilon$, $P \ll m$, $m/r \ll 1$)

$$\frac{dP(r)}{dr} = -\frac{\varepsilon(r)\, m(r)}{r^2}, \tag{7.21}$$

and is known as the equation of classical hydrostatic equilibrium. This equation describes the pressure gradient inside of a spherically symmetric mass distribution. The mass contained in a spherical shell of radius $r$ is given by

$$m(r) = 4\pi \int_0^r r'^2 \varepsilon(r') dr'. \tag{7.22}$$

The total mass $M$ of the star is given by $M = m(R)$, where $R$ denotes the stellar radius defined by $P(r = R) = 0$. The equation of hydrostatic equilibrium can be solved for a given equation of state (EoS) describing the matter (particle composition) of the star. In this assignment, we will study stars composed entirely of a relativistic gas of free quarks. The EoS of such a system is know as the MIT bag model equation of state [5]. It is given by

$$P = (\varepsilon - 4B)/3. \tag{7.23}$$

Here, $P$ and $\varepsilon$ denote pressure and energy density in units of MeV fm$^{-3}$, and $B = 57$ MeV fm$^{-3}$ is a constant.

**Tasks**

Write a Fortran program that solves the coupled set of equations (7.20) and (7.22) numerically via one or more of the methods described in sections 4.8.1 to 4.8.3 for given central energy densities $\varepsilon_c \equiv \varepsilon(r = 0)$ ranging from 4.2 $B$ to 2000 $B$. The finite-difference representation of equations (7.20) and (7.22) is given by

$$\Delta m = 4\pi \varepsilon r^2 \Delta r, \tag{7.24}$$

$$P(r + \Delta r) = P(r) - f(r)\Delta r, \tag{7.25}$$

where

$$f(r) \equiv \frac{(\varepsilon + P)(4\pi r^3 P + m)\kappa}{r^2(1 - 2m\kappa/r)}. \tag{7.26}$$

The units in equations (7.24) to (7.26) are as follows: $[\varepsilon] = $ MeV/fm$^3$, $[P] = $ MeV/fm$^3$, $[m] = $ MeV, $[r] = $ fm.

In the units of $G = c = 1$, the mass of the Sun is $M_\odot = 1.47$ km. On the other hand $M_\odot c^2 = M_\odot = 1.115\,829 \times 10^{60}$ MeV so that $\kappa \equiv 1.475 \times 10^{18}\text{fm}/1.115\,829 \times 10^{60}\text{MeV}$, which relates MeV to fm.

**Program design:**
1. Choose a step size of $\Delta\varepsilon_c = B/10$ and a radial step size of $r = 10$ meters.
2. Use the DO WHILE construct for the range of central densities ($\Delta\varepsilon_c$) and for the condition of $P(r) > 0$.
3. The stellar radius $R$ (in km) and mass $M$ (in units of the mass of the Sun, $M_\odot$) are to be written, for each central density $\varepsilon_c$, to an external data file.
4. Generate plots for $M/M_\odot$ as a function of $R$. Repeat the calculation for Newtonian stars (i.e., equation (7.21)) and show the outcome in the same plot. Comment on any differences.

**Additional tasks:**

Have your program also compute the gravitational redshift of light,

$$z = \left(1 - \frac{2\,M}{R}\right)^{-1/2} - 1,$$

as a function of mass $M$ and write the results to an external data file. Have your program generate a plot of $z$ as a function of $M/M_\odot$. Repeat the calculation for Newtonian stars and show the outcome in the same plot. Comment on any differences.

## 7.11 Proton in constant electric and magnetic fields

The non-relativistic equation of motion of an electric charge $q$ with mass $m$ moving in a combined electric and magnetic field is given by

$$\frac{d\vec{v}}{dt} = \frac{q}{m}\vec{E} + \frac{q}{m}\vec{v} \times \vec{B}.$$

The following conditions will produce a trochoidal motion in the $x - y$ plane. Let

$$\vec{B} = B_z\vec{k}, \quad \vec{E} = E_y\vec{j}, \quad \vec{v}(0) = v_y(0)\vec{j}, \quad \vec{r}(0) = 0\vec{i} + 0\vec{j}.$$

Then, the $x$ and $y$ components of the acceleration are given by

$$\frac{d^2x}{dt^2} = \frac{q}{m}\frac{dy}{dt}B_z, \tag{7.27}$$

$$\frac{d^2y}{dt^2} = \frac{q}{m}E_y - \frac{q}{m}\frac{dx}{dt}B_z. \tag{7.28}$$

equations (7.27) and (7.28) can be written as a system of coupled first-order differential equations,

$$\dot{x} = v_x, \tag{7.29}$$

$$\dot{v}_x = \ddot{x}, \tag{7.30}$$

$$\dot{y} = v_y, \tag{7.31}$$

$$\dot{v}_y = \ddot{y}. \tag{7.32}$$

The input data and initial conditions for this problem are $m = 1.0 \times 10^{-27}$ kg, $q = 1.0 \times 10^{-19}$ C, $E_y = 1.5 \times 10^6$ V m$^{-1}$, $B_z = 0.1$ T, $x(0) = 0$, $y(0) = 0$, $v_x(0) = 0$, $v_y(0) = 1.0 \times 10^6$ m s$^{-1}$.

**Tasks:**

Write a structured Fortran program which solves equations (7.29) to (7.32) numerically via one or more of the methods described in sections 4.8.1 to 4.8.3 for $0 < t < t_{\text{final}}$, where $t_{\text{final}} = 1.2 \times 10^{-6}$ s.

**Program design:**

1. Use a temporal step size of $\Delta t = 5.0 \times 10^{-10}$ s.
2. Assign numerical values to $m$, $q$, $E_y$, $B_z$ $x(0)$, $y(0)$, $v_x(0)$, and $v_y(0)$ in a SUBROUTINE named input_data.
3. Define a SUBROUTINE named write which writes the values assigned to $m$, $q$, $E_y$, $B_z$ $x(0)$, $y(0)$, $v_x(0)$, and $v_y(0)$ back to standard output.
4. The differential equations (7.29) to (7.32) are to be solved in a SUBROUTINE named diffeqs.
5. The proton's position, i.e., $y$ as a function of $x$ (given in meter), is to be written to an output file.
6. Have your program generate a plot which illustrates the proton's position graphically.

## 7.12 Square voltage pulse applied to a RC circuit

A resistor-capacitor circuit (RC circuit) is an electric circuit composed of resistors (R) and capacitors (C) driven by a voltage or current source [1, 2]. An RC circuit consisting of only one resistor and one capacitor is shown in figure 7.2. A square voltage pulse with a time dependence given by

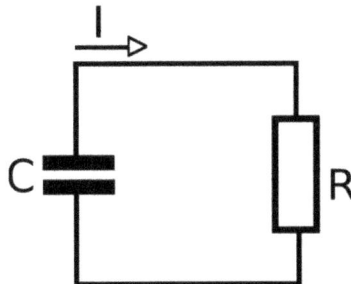

**Figure 7.2.** Illustration of a simple RC circuit.

$$U(t) = \begin{cases} 0, & \text{if } t < t_1 \\ U_0 = 6\,\text{V}, & \text{if } t_1 \leqslant t \leqslant t_2 \\ 0, & \text{if } t_2 < t \end{cases} \tag{7.33}$$

is applied to the RC circuit shown in figure 7.2. The circuit current $\dot{q}$ ($=I$) is given by the equation

$$R\frac{dq}{dt} + \frac{1}{C}q = U(t), \tag{7.34}$$

where $R = 100\ \Omega$ and $C = 4.7 \times 10^{-5}$ A s $\text{V}^{-1}$. The initial condition is $q(0) = 0$.

**Tasks:**

Write a complete Fortran program which solves equation (7.34) (i.e., computes $q(t)$ and $I(t)$) numerically via one or more of the methods described in sections 4.8.1 to 4.8.3 for times $0 < t < t_{\text{final}}$, where $t_{\text{final}} = 4RC$, $t_1 = 0$, and $t_2 = 2RC$.

**Program design:**

1. There is NO input from keyboard.
2. Use $\Delta t = RC/200$ for the incremental time step.
3. The results for $q(t)$ and $I(t)$ are to be written to external data files.
4. The mathematical solution of equation (7.34) for the current $I(t)$ is given by

$$\bar{I}(t) = \frac{U_0}{R}(e^{-(t-t_1)/(RC)}\,\Theta(t - t_1) - e^{-(t-t_2)/(RC)}\,\Theta(t - t_2)), \tag{7.35}$$

where $\Theta$ denotes the Heaviside step function given by

$$\Theta(x) = \begin{cases} 0, & \text{if } x < 0 \\ 1, & \text{if } x \geqslant 0 \end{cases}$$

Design your code such that $\bar{I}(t)$ is computed for $0 < t < t_{\text{final}}$. The result is to be written to an external data file.

5. Generate a plot which shows $I(t)$ and $\bar{I}(t)$ for $0 < t < t_{\text{final}}$.

## 7.13 Mutual inductance of two coils

Mutual inductance is the basic operating principal of the transformer, motors, generators, and any other electrical component that interacts with another magnetic field. Mutual induction is defined as the current flowing in one coil that induces an current in an adjacent coil. An example is shown in figure 7.3, where current $I_1$ flowing in coil $L_1$ caused a current $I_2$. The differential equations which describes the flow of the electric currents are given by

$$L_1\,\dot{I}_1 + R_1\,I_1 - L_{12}\,\dot{I}_2 = U, \tag{7.36}$$

$$L_2\,\dot{I}_2 + R_2\,I_2 - L_{12}\,\dot{I}_1 = 0, \tag{7.37}$$

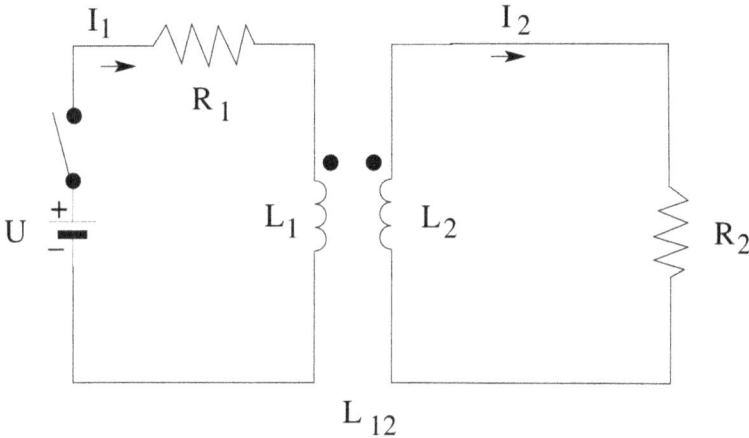

**Figure 7.3.** Mutual inductance between coils.

where $L_1 = 1.6$ H, $L_2 = 0.9$ H, $L_{12} = 0.72$ H, $R_1 = 48\ \Omega$, $R_2 = 27\ \Omega$, and $U = 240$ V[3].

**Tasks:**

Write a structured Fortran program which solves equations (7.36) and (7.37) numerically via one or more of the methods described in sections 4.8.1 to 4.8.3 for times $0 \leqslant t \leqslant t_{final}$, where $t_{final} = 0.25$ seconds. The initial conditions for the electric currents are $I_1(0) = I_2(0) = 0$ A.

**Program design:**

1. Choose $\Delta t = t_{final}/250$ for the temporal step size.
2. The analytic solutions of equations (7.36) and (7.37) are given by

$$\bar{I}_1(t) = 5 - 2.5\,e^{-75t} - 2.5\,e^{-75t/4} \tag{7.38}$$

$$\bar{I}_2(t) = -3.33\,e^{-75t} + 3.33\,e^{-75t/4}. \tag{7.39}$$

   Use FUNCTIONs to compute the analytic solutions given by equations (7.38) and (7.39).
3. Use the Fortran module feature to assign values to $L_1$, $L_{12}$, $L_2$, $R_1$, $R_2$, and $U$.
4. The initial value of $\bar{I}_1$ and $\bar{I}_2$ are keyboard input.
5. Generate a plot which shows the numerical as well as the analytic results for $I_1(t)$ and $I_2(t)$ for $0 \leqslant t \leqslant t_{final}$.

## 7.14 The accelerating Universe

The Friedman equation is a linear second-order differential equation that predicts the size of our Universe as a function of cosmic time. It is given by

---

[3] H=$\Omega$ s, F=s/$\Omega$.

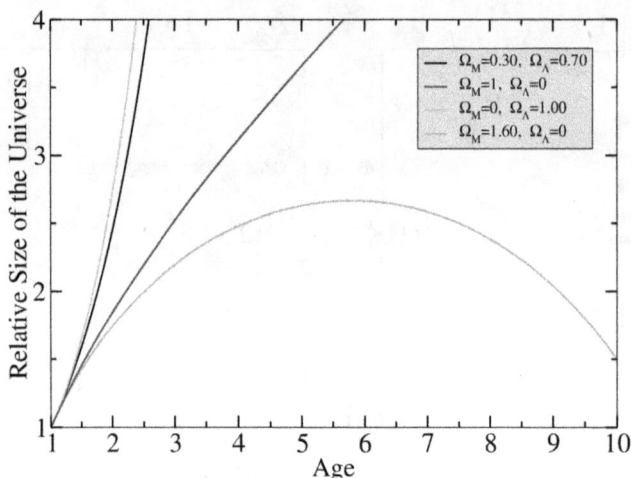

**Figure 7.4.** Solutions of the Friedman equation.

$$\ddot{R}(t) = -\frac{\Omega_M}{2R(t)^2} + \Omega_\Lambda\, R(t), \tag{7.40}$$

where the input parameters are the matter density, $\Omega_M$, and the cosmological constant (or vacuum density), $\Omega_\Lambda$. Current estimates of the two densities are in the range 0.31–0.33 for matter density and 0.70–0.72 for vacuum density. The numbers add up to almost exactly 1.0. Lately, data has been tending toward a sum of 1.02. The solution of the Friedman equation depends critically on $\Omega_M$ and $\Omega_\Lambda$. To solve this equation, one starts at the present where $R = 1$ and $\frac{dR}{dt} = 1$. Solutions of the Friedman equation for different combinations of $\Omega_M$ and $\Omega_\Lambda$ are shown in figure 7.4.

**Tasks:**

Write a Fortran program that solves equation (7.40) from the present time ($t = 1$) to 10 times the current age ($T = 10$) of the Universe, i.e., $1 \leqslant t \leqslant T = 10$. For this purpose, write equation (7.40) as a system of two coupled first-order differential equations. These equations are then solved using the numerical Euler forward scheme.

**Program design:**

1. Include a short preamble at the beginning of the program.
2. The user should be prompted (keyboard input) to enter values for $\Omega_M$ and $\Omega_\Lambda$. Display these values on the screen.
3. The Euler forward scheme is to be used to solve the coupled system of differential equations. Use a temporal step size of $\Delta t = 0.001$.
4. The (relative) size of the Universe $R(t)$ as a function of time $t$ is to be written to an output file.
5. The code should be run for the following combinations of matter and vacuum densities:

$(\Omega_M, \Omega_\Lambda) = (0.30, 0.70)$, which is approximately the measured actual values.

$(\Omega_M, \Omega_\Lambda) = (1.0, 0.0)$, which represents the present total density but with no vacuum density contribution.

$(\Omega_M, \Omega_\Lambda) = (0.0, 1.0)$, which represents no matter density and only vacuum density.

$(\Omega_M, \Omega_\Lambda) = (1.60, 0.0)$, which corresponds to a Universe that has 1.6 times the mass density of our Universe but with no vacuum density contribution.

6. The results are to be shown graphically.

## 7.15 An economic demand-and-supply model

In this assignment, we explore the dynamics of a basic economic demand-and-supply model, encapsulated in a differential equation framework. This model elucidates the relationship between the price $P(t)$, supply $S(t)$, and demand $D(t)$ of a commodity over time. By studying the interplay of these variables, we aim to explore the mechanisms that drive price fluctuations in markets.

A basic economic demand-and-supply model is represented by the differential equation

$$\frac{dP}{dt} = k(D(t) - S(t)), \tag{7.41}$$

where $P(t)$ denotes the price, $S(t)$ represents the supply, and $D(t)$ stands for the demand of a commodity over time $t$. The demand $D(t)$ and supply $S(t)$ are given by

$$D(t) = a - b\, P(t), \tag{7.42}$$

$$S(t) = c\, (1 - \cos(\alpha t)), \tag{7.43}$$

where $k = 5.5$ is a constant, and the values of $a$, $b$, $c$, and $\alpha$ are 3.67, 0.066, 2.03, and 0.77 respectively. If demand surpasses supply, $dP/dt > 0$, resulting in a price increase. Conversely, if supply exceeds demand, $dP/dt < 0$, leading to a price decrease.

The equation (7.41) can be solved analytically, yielding

$$P(t) = \left( P(0) - \frac{a-c}{b} - \frac{k^2 bc}{k^2 b^2 + \alpha^2} \right) e^{-kbt} + \frac{a-c}{b}$$
$$+ \frac{kc}{k^2 b^2 + \alpha^2}(kb\cos(\alpha t) + \alpha \sin(\alpha t)). \tag{7.44}$$

**Tasks:**

Create a structured and well-commented Fortran program to numerically solve the first-order differential equation (7.41) for $0 \leqslant t \leqslant T_{max} = 40$, with the initial condition $P(0) = 1.0$.

**Code design:**

1. The program must contain a preamble.
2. The program must prompt the user to input the time-step value $\Delta t = 0.01$ from the keyboard. Use the `advance='NO'` option with the `WRITE` statement to prevent the output from advancing to the next line after writing.
3. Use a named `DO WHILE` loop to compute $P(t)$, $D(t)$, and $S(t)$ for $0 \leqslant t \leqslant T_{max}$.
4. Write the results for $P(t)$, $D(t)$, and $S(t)$ to output files named `P_t.dat`, `D_t.dat`, and `S_t.dat`, respectively.
5. Compute the analytic solution $P(t)$ using equation (7.44) and write the results to an output file named `P_analytic.dat`
6. Using the Python script `hw7_compare.py`, generate a plot comparing the numerical results for $P(t)$ with the analytical ones.
7. Using the Python script `hw7_DS.py`, create a second plot showing $D(t)$ and $S(t)$ for $0 \leqslant t \leqslant T_{max}$.
8. Use the `iostat` parameter option to capture the status of the `OPEN` operations. If an error occurs during the opening operations, display an appropriate error message on the terminal and terminate execution.

## 7.16 Photo-pion production in the Universe and the GZK cutoff

The Greisen–Zatsepin–Kuzmin (GZK) cutoff is an upper limit on the energy of cosmic rays that can propagate over long distances through the Universe without significant attenuation due to interactions with photons of the cosmic microwave background radiation. It arises from the process of photo-pion ($\gamma$–$\pi$) production, where high-energy cosmic ray protons ($p$) collide with low-energy photons ($\gamma$) from the cosmic microwave background and produce pions ($\pi$), which then decay into secondary particles. The scattering process can be illustrated as:

$$p + \gamma \longrightarrow n + \pi.$$

The energy of the incoming proton, $E_p$, is related to the photon energy, $E_\gamma$, the neutron mass, $m_n$, and the pion mass, $m_\pi$, as

$$E_p = \frac{m_n \, m_\pi (1 + m_\pi/(2m_n))}{2E_\gamma}, \tag{7.45}$$

where $E_p = 3 \times 10^{14}$ MeV, $m_n = 940.6$ MeV, $m_p = 938.27$ MeV, and $E_\gamma = 2.33 \times 10^{-10}$ MeV.

**Tasks:**

Write a structured and well-commented Fortran program which uses Newton's numerical root-finding method to determine the pion mass $m_\pi$ form equation (7.45).

**Program design**

1. Use an initial value of $m_\pi = 400$ MeV to start the root-finding algorithm. This value is entered via keyboard input. Use the `advance='no'` option in the `write` statement so that the cursor does not advance after writing.
2. Limit the maximum number of allowed iterations to 50.

3. Terminate the Newton root-finding scheme if the difference $\Delta \equiv |m_\pi^{(i+1)} - m_\pi^{(i)}| \leqslant \varepsilon$ between two successive iteration steps $i$ and $i + 1$, where the tolerance $\varepsilon = 5 \times 10^{-5}$ MeV.
4. Write a warning message to the screen if the root has not been found after the maximum number of allowed iterations.
5. Use FUNCTIONs to compute the values of $F(m_\pi)$, defined as

$$F(m_\pi) \equiv E_p - \frac{m_n\, m_\pi (1 + m_\pi/(2m_n))}{2E_\gamma}, \tag{7.46}$$

and its derivative $F'(m_\pi) = dF(m_\pi)/dm_\pi$.
6. Use the intent descriptor to declare the arguments in the FUNCTION subprograms.
7. For each iteration, write the values of the iteration index it, $\Delta$, and $m_\pi$ to standard output. The terminal dialog should be as shown below:

```
Starting value for m_pi (= 400 MeV): 400
it = 1 Delta = 0.235 697E+03   m_pion = 164.30 MeV
it = 2 Delta = 0.252 407E+02 m_pion = 139.06 MeV
  . . .
  . . .
  . . .
```

Note: The figures produced by your machine may vary slightly because of machine precision.
8. The program should compute the Lorentz factor of protons traveling at an energy $E_p = 3 \times 10^{14}$ MeV, which is given by $\gamma = E_p/m_p + 1$, and their speed in units of the speed of light ($c$), given by $v/c = (1 - 1/\gamma^2)^{1/2}$. The results are to be written to standard output as:

```
The Lorentz factor of the protons is: . . .
The speed of the protons v/c is: . . .
```

# References

[1] Edminister J and Nahvi M 2013 *Schaum's Outline of Electromagnetics* 4th edn (Schaum's Outline Series) (New York: McGraw-Hill Education)
[2] Edminister J and Nahvi M 2011 *Schaum's Outline of Electric Circuits* 5th edn (Schaum's Outline Series) (New York: McGraw-Hill Education)
[3] Oppenheimer J R and Volkoff G M 1939 On massive neutron cores *Phys. Rev.* **55** 374–81
[4] Tolman R C 1939 Static solutions of Einstein's field equations for spheres of fluid *Phys. Rev.* **55** 364–73
[5] Chodos A, Jaffe R L, Johnson K, Thorn C B and Weisskopf V F 1974 New extended model of hadrons *Phys. Rev.* D **9** 3471–95

**IOP** Publishing

# Introduction to Computational Physics for Undergraduates (Second Edition)

**Omair Zubairi and Fridolin Weber**

# Appendix A

## Cubic spline Fortran code

The Fortran code shown below solves the cubic spline equations presented in equations (4.24)–(4.33).

```fortran
program cubic_spline_interpolation
  implicit none

  integer, parameter :: n = 10
  real, dimension(n) :: x_data = [0.0, 0.5, 1.0, 1.5, 2.0,
2.5, 3.0&
  &, 3.5, 4.0, 4.5]
  real, dimension(n) :: y_data = [0.0, 1.0, 0.5, 2.0, 1.5,
0.7, 1.2&
  &, 2.5, 1.8, 1.0]
  integer, parameter :: m = 500
  real, dimension(m) :: x_interp, y_interp
  real :: h(n-1), a(n), b(n-1), c(n), d(n-1), alpha(n-1),
l(n),&
  & mu(n), z(n)
  integer :: i, j

  ! Initialize x_interp for interpolation
  do i = 1, m
  x_interp(i) = x_data(1) + (x_data(n) - x_data(1)) * (i-
1) / &
  & (m-1)
  end do

  ! Step 1: Compute the intervals h_i and coefficients a_i
```

```fortran
do i = 1, n-1
h(i) = x_data(i+1) - x_data(i)
a(i) = y_data(i)
end do
a(n) = y_data(n)

! Step 2: Compute the coefficients alpha_i
do i = 2, n-1
alpha(i) = (3.0/h(i)) * (a(i+1) - a(i)) - (3.0/h(i-1)) *
&
& (a(i) - a(i-1))
end do

! Step 3: Solve the tridiagonal system
l(1) = 1.0
mu(1) = 0.0
z(1) = 0.0
do i = 2, n-1
l(i) = 2.0 * (x_data(i+1) - x_data(i-1)) - h(i-1) * mu
(i-1)
mu(i) = h(i) / l(i)
z(i) = (alpha(i) - h(i-1) * z(i-1)) / l(i)
end do
l(n) = 1.0
z(n) = 0.0
c(n) = 0.0

do j = n-1, 1, -1
c(j) = z(j) - mu(j) * c(j+1)
b(j) = (a(j+1) - a(j)) / h(j) - h(j) * (c(j+1) + 2.0 * c
(j)) &
&/ 3.0
d(j) = (c(j+1) - c(j)) / (3.0 * h(j))
end do

! Step 4: Evaluate the spline at the interpolated points
do i = 1, m
do j = 1, n-1
if (x_interp(i) >= x_data(j) .and. x_interp(i) <=&
& x_data(j+1)) then
y_interp(i) = a(j) + b(j) * (x_interp(i) - x_data(j))&
& + c(j) * (x_interp(i) - x_data(j))**2 + d(j) *&
& (x_interp(i) - x_data(j))**3
```

```
end if
end do
end do

! Write the interpolated values to a file
 open(unit=10,  file='interpolated_values.dat',  status
='unknown')
do i = 1, m
write(10, *) x_interp(i), y_interp(i)
end do

! Write the original data to a file
   open(unit=20,    file='original_data.dat',    status
='unknown')
do i = 1, 10
write(20,*) x_data(i), y_data(i)
end do

close(10); close(20)

end program cubic_spline_interpolation
```

**IOP** Publishing

Introduction to Computational Physics for Undergraduates
(Second Edition)

Omair Zubairi and Fridolin Weber

# Appendix B

## Summary of modern Fortran features

Fortran 90/95/2000 introduces many advanced features that enhance the language's flexibility, efficiency, and ease of use. These improvements include:

- Dynamic memory management: Fortran 90 introduced the ability to allocate and deallocate memory dynamically, a feature that was not available in Fortran 77. This capability is crucial for developing efficient programs that handle varying amounts of data.

- User-defined data structures: The introduction of derived types in Fortran 90 allows users to define complex data structures. This feature aligns Fortran more closely with modern programming practices and languages, such as C++.

- Matrix operations: Fortran 90/95/2000 includes intrinsic support for matrix and array operations, enabling more concise and readable code for scientific and engineering computations. These operations are performed efficiently, leveraging optimized libraries.

- Operator definition and overloading: Modern Fortran allows users to define and overload operators, providing greater flexibility and making code more intuitive. This feature is particularly useful in numerical and algebraic computations.

- Intrinsics for vector and parallel processors: Fortran 90/95/2000 includes intrinsic functions designed for vector and parallel processing, allowing programmers to write code that takes full advantage of modern hardware architectures.

doi:10.1088/978-0-7503-6493-5ch9

- Object-oriented programming (OOP): Fortran 2003 introduced features that support OOP, including type extension, polymorphism, and dynamic type allocation. These features enable the creation of more modular and maintainable code.

- Enhanced control structures: Modern Fortran includes enhanced control structures, such as 'select case' and 'do while', which improve code readability and control flow management.

- Modules: Fortran 90 introduced modules, which help in organizing code, encapsulating data, and procedures. Modules also facilitate code reuse and better namespace management.

These enhancements make Fortran 90/95/2000 powerful tools for scientific and engineering programming, providing capabilities that were not possible with Fortran 77. Tables B.1, B.2, and B.3 serve as a condensed quick reference guide for programming in Fortran 90/95/2000. They also help in understanding and maintaining programs developed by others.

**Table B.1.** This table shows array operations in programming constructs.

| Description | Equation | F90/F95/2000 Operation |
|---|---|---|
| Scalar plus scalar | $c = a \pm b$ | $c = a \pm b$ |
| Element plus scalar | $c_{jk} = a_{jk} \pm b$ | $c = a \pm b$ |
| Element plus element | $c_{jk} = a_{jk} \pm b_{jk}$ | $c = a \pm b$ |
| Scalar times scalar | $c = a \times b$ | $c = a*b$ |
| Element times scalar | $c_{jk} = a_{jk} \times b$ | $c = a*b$ |
| Element times element | $c_{jk} = a_{jk} \times b_{jk}$ | $c = a*b$ |
| Scalar divide scalar | $c = a/b$ | $c = a/b$ |
| Scalar divide element | $c_{jk} = a_{jk}/b$ | $c = a/b$ |
| Element divide element | $c_{jk} = a_{jk}/b_{jk}$ | $c = a/b$ |
| Scalar power scalar | $c = a^b$ | $c = a**b$ |
| Element power scalar | $c_{jk} = a_{jk}^{b}$ | $c = a**b$ |
| Element power element | $c_{jk} = a_{jk}^{b_{jk}}$ | $c = a**b$ |
| Matrix transpose | $C_{kj} = A_{jk}$ | $C=\texttt{transpose}(A)$ |
| Matrix times matrix | $C_{ij} = \sum_k A_{ik} B_{kj}$ | $C=\texttt{matmul}(A, B)$ |
| Vector dot vector | $c = \sum_k A_k B_k$ | $c=\texttt{sum}(A*B)$ |
| | | $c=\texttt{dot\_product}(A, B)$ |

**Table B.2.** Fortran features to include intrinsic data types, relational operators, and flow control statement.

| Description | F77 | F90/F95/2000 |
| --- | --- | --- |
| Comment syntax | C,* | ! |
| Byte | character | character:: |
| Integer | integer | integer:: |
| Single precision | real | real:: |
| Double precision | double precision | real*8:: |
| Complex | complex | complex:: |
| Argument | parameter | parameter:: |
| Pointer | — | pointer:: |
| Structure | — | type:: |
| Equal to | .EQ. | == |
| Not equal to | .NE. | /= |
| Less than | .LT. | < |
| Less or equal | .LE. | <= |
| Greater than | .GT. | > |
| Greater or equal | .GE. | >= |
| Logical NOT | .NOT. | .NOT. |
| Logical AND | .AND. | .AND. |
| Logical inclusive OR | .OR. | .OR. |
| Logical exclusive OR | .XOR. | .XOR. |
| Logical equivalent | .EQV. | .EQV. |
| Logical not equivalent | .NEQV. | .NEQV. |
| Conditionally execute statements | if | if |
| | end if | end if |
| Loop a specific number of times | do # k=1,n | do k=1,n |
| | # continue | end do |
| Loop an indefinite number of times | — | do while |
| | — | end do |
| Terminate and exit loop | go to | exit |
| Skip a cycle of loop | go to | cycle |
| Display message and abort | stop | stop |
| Return to invoking function | return | return |
| Conditional array action | — | where |
| Conditional alternative statements | else | else |
| | elseif | elseif |
| Conditional array alternatives | — | elsewhere |
| Conditional case selections | if | select case |
| | end if | end select |

**Table B.3.** Overview of Fortran 90 intrinsic functions. The names of the arguments specify their type (i.e., X=Real, DX=Double precision, IX=Integer, Z=Complex).

| F90 | Function Type | Definition |
|---|---|---|
| SQRT(X) | Real | $\sqrt{X}$ |
| DSQR(DX) | Double precision | $\sqrt{DX}$ |
| ABS(X) | Real | $|X|$ |
| EXP(X) | Real | $e^X$ |
| DEXP(DX) | Double precision | $e^{DX}$ |
| LOG(X) | Real | $\log_e X$ |
| LOG10(X) | Real | $\log_{10} X$ |
| IFIX(X) | Integer | Truncate $X$ to an integer |
| AINT(X) | Real | Round number |
| NINT(X) | Real | Round $X$ to an integer |
| FLOAT(X) | Real | Converts $IX$ to real value |
| CEILING(X) | Real | Smallest integer $> X$ |
| FLOOR(X) | Real | Largest integer $< X$ |
| MOD(X,Y) | Real | Division remainder |
| CONJ(Z) | Real | Complex conjugate |
| IMAG(Z) | Real | Imaginary part |
| DBLE(X) | Double precision | Convert $X$ to double precision |
| AMAX1(X,Y,...) | Real | Maximum of $(X, Y, ...)$ |
| AMAX0(IX,IY,...) | Real | Maximum of $(IX, IY, ...)$ |
| AMIN0(IX,IY,...) | Real | Minimum of $(IX, IY, ...)$ |
| AMIN1(X,Y,...) | Real | Minimum of $(X, Y, ...)$ |
| MIN0(IX,IY,...) | Integer | Minimum of $(IX, IY, ...)$ |
| SIN(X) | Real | $\sin(X)$ |
| COS(X) | Real | $\cos(X)$ |
| TAN(X) | Real | $\tan(X)$ |
| ASIN(X) | Real | $\arcsin(X)$ |
| ACOS(X) | Real | $\arccos(X)$ |
| ATAN(X) | Real | $\arctan(X)$ |
| SINH(X) | Real | $\sinh(X)$ |
| COSH(X) | Real | $\cosh(X)$ |
| TANH(X) | Real | $\tanh(X)$ |

**IOP** Publishing

Introduction to Computational Physics for Undergraduates
(Second Edition)

Omair Zubairi and Fridolin Weber

# Appendix C

## Plotting using Python

There are many plotting programs and software available to use. Some are proprietary, such as MATLAB and tableau, while others are open source, such as Python and GNU plot.

In this appendix, we show how to plot data using the Python programming language. Python is an open source interpreted (compiles as it runs) language used for a plethora of applications, including high level data analysis and visualization.

Plotting data using Python is very useful and versatile and can be performed on multiple platforms Python is an object-oriented type language; therefore, it is easily integratable with other languages, such as FORTRAN and MATLAB.

The file extension for a Python program is `.py` – (i.e., `filename.py`). To compile and run a Python program (on a Linux/Unix machine via a terminal window) simply type

```
> python filename.py                                        ↵
```

Note, you may need the full version extension depending on your configuration. For example, if you have Python version 3.12 installed, your compile and run command would be

```
> python3.12 filename.py                                    ↵
```

doi:10.1088/978-0-7503-6493-5ch10     C-1    

The first command assumes you have configured your system to where you have created an alias for your Python install name in your bash file. Thus, the name Python is just pointing (referencing) Python3.12.

Figure C.1 shows a simple graph of some numerical data. The sample Python code which produced figure C.1 is shown below for your reference. This code reads data from a file, extracts specific columns, plots the data with customized axis limits, and labels and then saves the plot in both PDF and EPS formats:

```python
import numpy as np
import matplotlib.pyplot as plt

# Import data
data = np.loadtxt('datafile.dat')

# Reading in two columns:
column1 = data[:, 0] # First column from data (column 1)
column2 = data[:, 1] # Second column from data (column 2)
plt.plot(column1,column2)

# Setting horizontal axis

plt.xlim(0,5)
# Setting vertical axis
plt.ylim(0,25)

# Label axis
plt.xlabel('x')
plt.ylabel('f(x)')

# Saving plot into .pdf and .eps formats}
plt.savefig('graph.pdf')
plt.savefig('graph.eps')
# Show graph in popup window
plt.show()
```

Next, we will walk through this Python code. The first two lines of the code

```python
import numpy as np
```

```python
import matplotlib.pyplot as plt
```

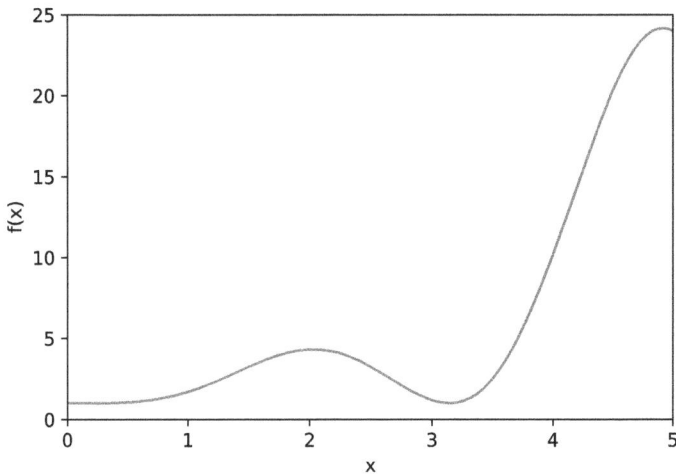

**Figure C.1.** Plot of the function $f(x) = 1 + x^2 \sin^2(x)$.

import the necessary libraries. numpy (imported as np) is used for numerical operations and matplotlib.pyplot (imported as plt) is used for plotting.

```
# Import data
```

```
data = np.loadtxt('datafile.dat')
```

This line reads data from a file named datafile.dat into a NumPy array called data. The np.loadtxt function assumes the data is in a text format. The next two lines of the code

```
# Reading in two columns:
```

```
column1 = data[:, 0] # First column from data (column 1)
```

```
column2 = data[:, 1] # Second column from data (column 2)
```

extract the first and second columns from the data array into variables column1 and column2, respectively. The ':' symbol selects all rows, and the '0' and '1' specify the first and second columns. A plot of column1 ($x$-axis) against column2 ($y$-axis) is generated by

```
plt.plot(column1, column2)
```

The next lines

```
# Setting horizontal axis
```

```
plt.xlim(0, 5)
```

```
# Setting vertical axis
```

```
plt.ylim(0, 25)
```

set the limits for the *x*-axis (0 to 5) and *y*-axis (0 to 25) The *x*-axis is labeled as 'x' and the *y*-axis as 'f(x)' with the following code:

```
# Label axis
```

```
plt.xlabel('x')
```

```
plt.ylabel('f(x)')
```

The lines

```
# Saving plot into .pdf and .eps formats
```

```
plt.savefig('graph.pdf')
```

```
plt.savefig('graph.eps')
```

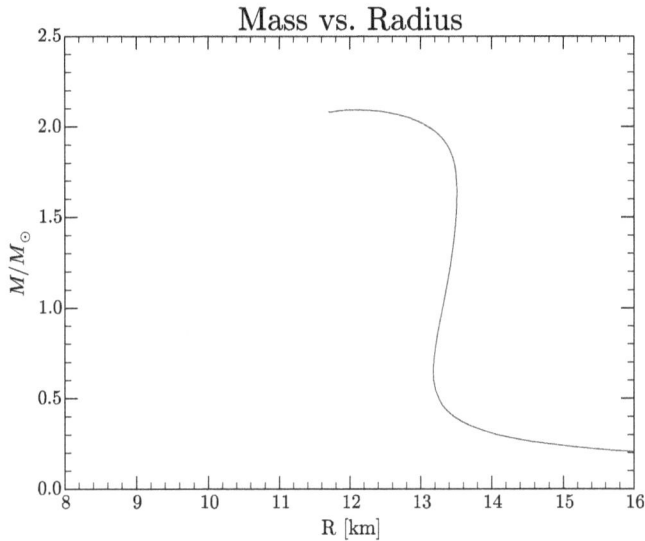

**Figure C.2.** Mass–radius plot for neutron stars.

save the plot in two formats: PDF and EPS, using the filenames `graph.pdf` and `graph.eps`. Finally, the plot is displayed in a popup window with the code

```
# Show graph in popup window



plt.show()
```

Note that the symbol '#' is used for comments. The sample code above covers the basics of plotting data. However, much more can be done using Python, such as configuring axes, including a title, inserting special symbols, etc. For example, in figure C.2, the axes are configured differently with more increments, there is a title for the graph, and a special character is inserted on the vertical axis.

The sample code below gives a more detailed description on how to reproduce figure C.2.

```
import numpy as np
import matplotlib.pyplot as plt

# Import data
mr = np.loadtxt('mass_radius.dat')

# Note:  The .dat file has 3 columns -- e_c, mass, radius
# but the columns are e_c = 0, mass = 1, radius = 2

# Specify which columns of data is being imported
plt.plot(mr[:,2],mr[:,1])

# Note:  ([x],[y])-->([radius], [mass]) = ([:,2], [:,1])

# Title of plot
plt.title('Mass vs. Radius', fontname='cmr10', fontsize=25)

# Configure ticks:
plt.minorticks_on()
plt.tick_params(axis='both', which='minor', length=5,
                width=0.75, labelsize=15)
plt.tick_params(axis='both', which='major', length=10,
                width=1.0, labelsize=15)

# Configure x-axis
plt.xlim((8,16))  #range of x-values
plt.xlabel(r'R [km]', fontname='cmr10', fontsize=16)
plt.xticks(fontname='cmr10', fontsize=15, size=15)


  # Configure y-axis
  plt.ylabel(r'$\mathrm{M/M_{\odot}}$', fontname='cmr10',
          fontsize=16)
  plt.yticks(fontname='cmr10', fontsize=15, size=15)

  # Save plot
  plt.savefig('mr.pdf')
  plt.savefig('mr.eps')
  plt.show()
```

This Python code performs the following tasks. First, the data are imported from the file mass_radius.dat into a NumPy array mr via

```
mr = np.loadtxt('mass_radius.dat')
```

The data file contains three columns: e_c, mass, and radius, mapped to indices '0', '1', and '2' respectively. The line

```
plt.plot(mr[:,2], mr[:,1])
```

plots the data where the *x*-axis represents radius and the *y*-axis represents mass. A title is added and the plot configured with

```
plt.title('Mass versus Radius', fontname='cmr10', font-
size=25)
plt.minorticks_on()
plt.tick_params(axis='both', which='minor', length=5,
width=0.75, labelsize=15)
plt.tick_params(axis='both', which='major', length=10,
width=1.0, labelsize=15)
```

This sets the title of the plot to 'Mass versus Radius' with a font size of 25; the minor and major ticks on both axes are configured with specified lengths, widths, and label sizes. The code for the axis configurations reads

```
plt.xlim((8,16)) # Range of x-values
plt.xlabel(r'R [km]', fontname='cmr10', fontsize=16)
plt.xticks(fontname='cmr10', fontsize=15, size=15)
plt.ylabel(r'M/M
_□', fontname='cmr10', fontsize=16)
plt.yticks(fontname='cmr10', fontsize=15, size=15)
```

This sets the limits of the *x*-axis from 8 to 16, labels the *x*-axis as 'R [km]' with 'cmr10' font and fontsize of 16, and configures *x*-axis tick labels with 'cmr10' font, fontsize of 15, and size of 15. Labels the *y*-axis with LaTeX formatting for 'M/M$_\odot$', ensuring proper rendering of the math symbols (M/M$_\odot$) with specified font ('cmr10') fontsize of axis ticks.

LaTeX is often integrated with Python plotting libraries such as Matplotlib to enhance the presentation of mathematical expressions and symbols. In a typical Python plot, LaTeX commands can be embedded directly within strings used for annotations such as axis labels, titles, and legends. For instance, in figure C.2, the code

```
plt.ylabel(r'$\mathrm{M/M_{\odot}}$', . . .)
```

was used to set the label of the $y$-axis. The string passed to `ylabel` is a raw string (indicated by the prefix r) that contains LaTeX commands. The raw string ensures that backslashes are treated as literal characters, which is necessary for LaTeX syntax. The dollar signs ($) delimit a LaTeX math mode environment, which is used for rendering mathematical expressions. The `\mathrm` command is used to ensure that the text $M/M_\odot$ is rendered in an upright (Roman) font, as opposed to the default italic font used in math mode, in which case $M/M_\odot$. The subscript '_' is used to denote that the character following it (`\odot`) should be a subscript on M. The `\odot` command is a LaTeX symbol representing the solar mass symbol ($\odot$).

**IOP** Publishing

# Introduction to Computational Physics for Undergraduates (Second Edition)

Omair Zubairi and Fridolin Weber

# Appendix D

# Fortran 90 sample program illustrating good programming

The following is a sample program which calculates the fraction of gas molecules in Venus's atmosphere that have speeds less than a given value, based on the Maxwell–Boltzmann distribution,

$$f(v; T) = 400\pi \left(\frac{m}{2\pi k_B T}\right)^{3/2} \int_0^v dv' \, v'^2 \, e^{-mv'^2/(2k_B T)}, \qquad (D.1)$$

where $f(v; T)$ is the fraction (between 0% and 100%) of molecules in a gas that have speeds less than $v$, at a given gas temperature $T$. The integral is computed numerically using the trapezoidal method with the integration formula given by equation (4.50). This program serves to illustrate good Fortran programming practices.

```fortran
!* module
   module constants
   implicit none
   real, parameter :: pi   = acos(-1.0)  ! Define pi
   real, parameter :: k_B  = 8.3145      ! k_B in Joule/mol/K
   real, parameter :: m_N2 = 0.028014    ! Molar mass of N2 in kg/mol
   real, parameter :: m_CO2= 0.04401     ! Molar mass of CO2 in kg/mol
   end module constants

!* main program
   program atmosphere

! Purpose: Calculate the fraction of gas molecules in Venus's
```

```
! atmosphere (uniform temperature T) whose speeds are less than a
! given speed, v.

! Given:
!   v     = Speed (m/sec)
!   m_N   = Molar mass of Nitrogen = 28.014 g/mol
!   m_CO2 = Molar mass of Carbon dioxide = 44.01 g/mol
!   Venus's atmosphere consists of 3.5% N2 and 96.5% CO2
!   k_B   = Boltzmann constant = 8.319 Joule/mol/K
!   T     = Temperature (K)

    use constants

    implicit none
    real ::    m, T_C, T_K, a, b, h, sumint, integral, vp, f_vT, f
    integer :: v, k, n=100, i

! open data file
    open(unit=100, file='atmosphere.dat', status='unknown', iostat=i)
    if(i /= 0) then
       write(*,*) 'Error opening of data file'
       stop
    end if

! read temperature from keyboard
    write(*, *) 'Enter temperature in Celsius: '
    read(*, *) T_C

! compute absolute temperature (in K)
    T_K = 273.15 + T_C

! compute average mass in kg/mol
    m = (m_N2*0.035 + m_CO2*0.965)

    a = 0.0
    speed_v: do v = 0, 3000, 10      ! Speeds v from 0 to 3000 m/s
                b = real(v)          ! Upper limit of integral (v)
                h = (b-a)/real(n)    ! Step size
                sumint = 0.0         ! Initialize summation variable

    trapezoidal: do k=1, n-1
```

```
             vp = a + h * real(k)
             sumint = sumint + f(vp,m,T_K)
          end do trapezoidal

       integral = h * ( f(a,m,T_K)+f(b,m,T_K)+2.0*sumint ) / 2.0

! compute Maxwell-Boltzmann distribution function f(v;T)
  f_vT = 4.0*pi * sqrt(m/(2.0*pi*k_B*T_K))**3 * integral

! write results to data file
  write (100, *) b, f_vT*100.0  ! output f_vt in %

  end do speed_v

  close(unit=100)

end program atmosphere


!* function sub-program
     function f(y,m,T)
     use constants, only: k_B
     implicit none
     real, intent (in) :: y, m, T
     real :: f

     f = y*y * exp(-m*y*y / (2.0*k_B*T))
     end function f
```

The various steps of this program are as follows:
1. Module declaration: The module named constants is declared at the beginning. Modules in Fortran are used to group related variables and procedures.
2. Implicit none: This statement ensures that all variables must be explicitly declared, helping to prevent errors due to undeclared variables.
3. Constants and definitions: Several constants ($\pi$, $k_B$, $m_{N_2}$, $m_{CO_2}$) are defined using the parameter attribute
4. Program declaration: The main program is named atmosphere.
5. Using the 'constants' Module: The use constants statement allows the main program to access the constants defined in the constants module.
6. Implicit none: This ensures that all variables must be explicitly declared.

7. Variable declarations: Various variables are declared for later use in the program, such as m, T_C, T_K, a, b, h, sumint, integral, vp, f_vT, f, v, k, n=100, and i.

8. Open data file: The program attempts to open a file named atmosphere.dat for writing results. If the file cannot be opened, an error message is displayed and the program stops.

9. User input: The program prompts the user to enter the temperature in Celsius, which is read and stored in the variable T_C.

10. Temperature conversion: The temperature in Celsius (T_C) is converted to Kelvin (T_K).

11. Molecular mass calculation: The average molecular mass of Venus's atmosphere, based on its composition (3.5% $N_2$, 96.5% $CO_2$), is calculated in $kg\,mol^{-1}$.

12. Integration setup: Variables a and b are set to '0' and the current speed v, respectively. The step size 'h' for the numerical integration is calculated. The variable sumint is initialized to '0' for summation.

13. Numerical integration: The program performs a numerical integration to compute the fraction of molecules with speeds less than v using the trapezoidal rule given by equation (4.50). The integral calculation is done in a loop where the function f (y, m, T) is evaluated and summed.

14. Maxwell–Boltzmann distribution: The Maxwell–Boltzmann distribution function $f(v; T)$ is calculated based on the integral result.

15. Output: The program writes the results (speed and fraction of molecules) to the data file atmosphere.dat.

16. Function program: The function f computes the integrand of equation (D.1) for input variables y, m, and T.

www.ingramcontent.com/pod-product-compliance
Lightning Source LLC
Chambersburg PA
CBHW080548220326
41599CB00032B/6402